Maclean

MINIATURE ROSES

MINIATURE ROSES

Dawn and Barry Eagle

Collins
AUCKLAND SYDNEY LONDON

ACKNOWLEDGEMENTS

We would like to thank all those who over the years
have given help and encouragement as we grew our
roses. In particular we would like to acknowledge
material from Harmon Saville of Nor' East Miniature
Roses, U.S.A.; William Warriner of Jackson and
Perkins, U.S.A.; Ralph Moore of Sequoia Nursery,
U.S.A.; Wilhelm Kordes of W. Kordes 'Sonne',
Germany; Patrick Dickson of Dickson Nurseries,
Northern Ireland; and Samuel McGredy of Auckland,
New Zealand, who provided comments and
illustrations for the book. All other photographs not
acknowledged were taken by Barry Eagle, mostly at
the nursery in Christchurch, New Zealand.

First published 1985
William Collins Publishers Ltd
P.O. Box 1, Auckland

© Dawn and Barry Eagle 1985

ISBN 0 00 217568 1

Typeset by Saba Graphics Ltd, Christchurch
Printed in Hong Kong by Dai Nippon Printing Co.
Design by Grant Nelson

CONTENTS

INTRODUCTION

My association with roses goes back a number of years. My father had been a keen gardener; I guess he must have had green fingers. He was constantly bringing home bits of this and that, pieces of plants he had been given, and they went into the few empty spaces that remained. Sometimes there was no real space, but they always went in somewhere. All through the garden were his roses. They didn't receive a lot of special attention. They must have been sprayed occasionally, but I also remember collecting buckets of warm soapy water after the family wash and dousing the roses with this to keep the greenfly in check. I also discovered two very important things: roses don't grow without water, and, if you cut them back after flowering, they will continue to bloom from spring to autumn.

When we were married Dawn said, 'No roses!' There were no roses in her bridal bouquet and we had no roses in the garden. For the first two years there was no opportunity to grow roses, even if we had wished to. We were renting a small cottage on a farm close to Putaruru in the centre of the North Island of New Zealand. As we were both working, gardening meant little more than keeping the lawn cut and the weeds down — and we soon found that a browsing sheep could do this job quite well.

When we returned to Christchurch to a home and garden of our own, it wasn't long before I came home from work one day to find that we now had half a dozen roses. What they were doesn't really matter. They were the start of our rose garden and over the next twenty years the original six bushes were to become more than 400 bush roses and another 400 miniature roses. But there were no miniature roses among those first plantings; they were to come later.

We were not impressed with our first contact with miniature roses. Knowing of our interest in roses, someone gave us a packet of 'Fairy' rose seed. Only a few grew. They had small, single, pink flowers on tiny weak plants and they soon perished — except one. It was a deeper pink and had a few more petals. Perhaps because it was a bit more interesting, it got better attention, living to be planted out into the garden where it grew into a small bush about 15 cm tall. Never very spectacular, it was eventually replaced with something better. We now know we were experiencing the usual trials and results of anyone who sets out to grow roses from seed.

A year or two later we bought a collection of miniature roses — 'Beauty Secret', 'Cinderella', 'Colibri', 'Jeanie Williams', 'Humpty Dumpty'. And they grew! They didn't stop at 15 cm but went on, some to 30 and 40 cm, growing without any special attention. They had true rose flowers. 'Beauty Secret' was the most attractive deep red rose that we had ever grown. The flowers were only about 2 cm across but they were, after all, 'miniature' roses. They were different. They didn't take much space. They were easy to grow. We bought some more. And then still more!

Barry Eagle

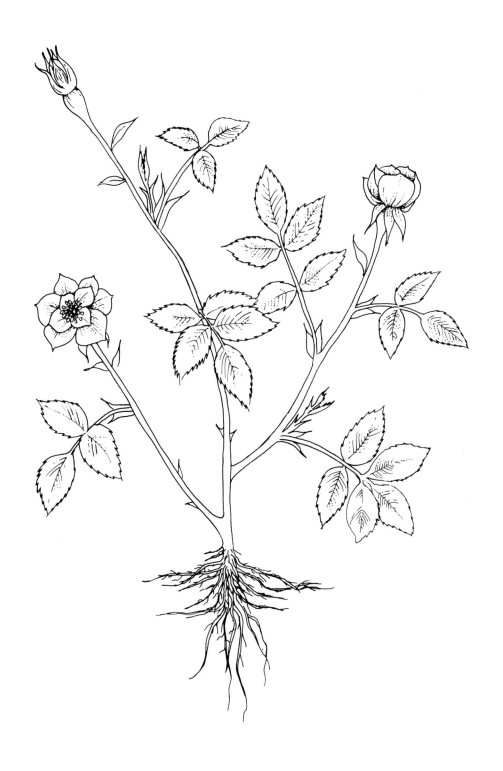

WHAT IS A MINIATURE ROSE?

When visitors come in to look at our roses and they see some of the miniatures growing 60 cm and more tall, their reaction is frequently, 'That can't be a miniature rose!' This is understandable. After all, a miniature is something small and for many people a miniature rose means small flowers on short stems growing on a small bush with small leaves. For others, only the size of the flower matters and varieties such as 'Climbing Cécile Brunner', which may span distances of up to 4 and 5 m, are considered miniature roses.

What, then, is a miniature rose? A few years ago I asked Ralph Moore of California this question and he replied, 'Miniature roses are descendants of *Rosa chinensis minima* (*Rosa rouletii*). They have tiny leaves, are much branched and make thick, compact plants. Buds, flowers, leaves, stem, thorns — every part of the plant is much reduced in size as compared to the larger forms of roses, yet in perfect proportion.'

It is this last property which, for us, determines a miniature rose. Miniature roses are scaled-down versions of larger roses, with every part — buds, flowers, leaves, stems, internodes, thorns — in perfect proportion. To understand miniature roses better, let us talk for a while about roses in general.

Modern garden bush roses come in a large variety of sizes and shapes, with a corresponding variation in the form, size and colour of the flowers. The typical rose bush grows about a metre high and, when in full growth, can easily spread 70 cm and more across, but there are just as many bush roses which do not fit these measurements. Some roses like 'Topsi' or 'Regensberg' grow only half this height. Others, such as 'Folklore' and 'Queen Elizabeth', reach for the sky and will easily grow 2 m and more. Some of these taller varieties are perhaps best thought of as shrub roses; we, though, cannot decide where bush roses finish and shrub roses begin. Why 'Fred Loads' is classified as a shrub and 'Eye Paint' as a bush, we cannot tell. For us they both grow about the same height, but it is 'Eye Paint' which is the more shrub-like with its shorter stems and masses of flower from top to bottom.

We all recognise climbing roses. Or do we? Most people who speak to us of climbing roses are thinking of a rose which produces long climbing shoots, or whips, which need to be tied and trained into the required form against a trellis or fence and may stretch up to 2 m in both directions. They forget about 'Handel' and other similar roses which are climbers, but of a more upright form. For many years we grew 'Handel' as an upright pillar, not much taller than some of our bush roses. Many admired the rose but few realised that it was a climber. What about ordinary climbers? A few years ago, while visiting rose-growing friends, we admired a length of trellis covered by a climbing rose with masses of deep red blooms; it was one of the best climbing roses we had seen for some time. On inquiry the rose turned out to be 'Uncle Walter' — and 'Uncle Walter' is a bush rose!

The confusion increases when we consider hybrid teas and floribundas. These names refer more to the type of flowers and how they grow than to the bush on which they are grown. Hybrid tea roses were developed late in the nineteenth century from a cross between the hybrid perpetuals and the tea roses. They combined the hardiness and vigour of the hybrid perpetuals with the elegance and colour range of the tea roses, and by early in the 1900s had largely replaced the roses from which they came. For many, it is this time, when hybrid teas were introduced, which separates 'old' roses from our more 'modern' roses; roses which originated before this time are 'old' roses. This does not mean that the 'old' rose bush growing in your garden has been growing all that time, but it does mean that the parent from which your bush and many others have come, dates from before the time of the hybrid teas. Hybrid tea roses typically have pointed buds which, as they open and unfurl, give a bloom with a pointed centre surrounded by a circular arrangement of petals. This is the shape which has come to be regarded as the classical rose form. Such roses grow with one flower on a long stem, occasionally with one or two side buds.

During the 1920s the Danish firm of Poulsen crossed a hybrid tea rose, 'Red Star', with the dwarf polyantha rose, 'Orleans Rose', to produce the first of what were to become the floribunda roses. The early floribunda roses frequently had fewer petals than the hybrid teas and the flower shape was more informal. More importantly, as the name suggests, the bushes produced an abundance of flowers. The stems were still long, but the flowers came in clusters, sometimes with twenty and more blooms in the same flower head. Over the following years floribundas were crossed with hybrid teas and the results crossed again with floribundas and hybrid teas. Because the resulting offspring show signs of both lines of parentage, it becomes impossible to draw clear lines between them. Some roses are definitely hybrid teas; others are undoubtedly floribundas — but there is a large group in between. To cover some of these in-between roses, the American rose growers introduced the term 'grandiflora', but this is not widely used outside the United States. More recently, someone coined the term 'floratea'. It certainly suggests a rose with characteristics of both groups, but the name has never been accepted.

For our gardens, we select roses to suit the situation and purpose we have in mind and we plant them, short or tall, hybrid tea or floribunda, in whatever place best serves our purpose. It really doesn't matter. We buy the rose we have seen and liked so much, and worry later about where it is to be planted. And when the blooms have been cut and placed in a vase in a position of honour, it doesn't matter whether they came from a bush or a climber. They are beautiful!

The same is true of miniature roses. When the flowers are picked, there is nothing to show whether the plant was short, tall or climbing. Provided the flower is small ('much reduced in size') and the leaves, stems, internodes, thorns — everything else — is in proportion, then it is a miniature rose. There are, however, still problems. Some roses are clearly miniature and others, just as obviously, are not, but there is a growing number of roses which, while not truly miniature, do not really belong in any other class. Many of the more recent varieties come into this in-between group. It is almost as though intro-

ducers have decided to give us 'bigger and better' miniature roses. With miniatures, however, bigger does not necessarily mean better. Just as in the past new classifications were made as the need arose, it could be time to have a new name for these in-between roses. 'Patio roses' and 'mini-flora' are names which have been suggested but, at this stage, none has received the universal nod of approval.

To overcome this problem of whether a rose is truly a miniature, this book will (unless otherwise noted) use the registered classification. If the introducer calls it a miniature, so will we.

Miniature roses have as much variation in form of growth and flower as their larger cousins. The smallest miniature roses grow only 15 to 20 cm tall. 'Si', said to be the smallest rose in commercial production, is the smallest of these. Our original plant, grown continuously in a pot over the past eight years, has never been more than 15 cm high. During this time it has grown into a sturdy little bush as wide as it is tall. Plants in the garden grow a little taller. Despite its size, 'Si', like all miniature roses, has flowers and foliage in perfect proportion. There are many other such varieties, including 'Bambino', 'Midget', 'Snow White' and 'Yellow Bantam'. These are sometimes referred to as 'micro minis' and grow a little taller than 'Si' but usually remain under 30 cm. Most of these are older varieties although a few, like 'Little Linda', are more recent introductions. All have short, fine leaves. Leaf size often gives an indication of the height that a miniature rose may reach; as a rule of thumb, the finer the leaf, the shorter the plant.

At the other end of the scale are miniature roses which, if left to their own devices, will readily grow to 1.5 m. Some of these are climbers but others have the characteristics of bush roses. As with most plants, if you starve them you stunt their growth, but when you treat them as they deserve they just want to grow tall. Such a rose is 'Swedish Doll'. With frequent trimming and no feeding, it is possible to keep it down to 60 or 70 cm. One year when it didn't get trimmed back and was left to its own devices, it peered happily over the top of a 1.5 m fence. And this is why so much confusion exists. How can a rose which grows so tall be a miniature rose? Remember that it is not the size of the bush which makes a miniature rose, but the size of the flowers and leaves. No matter how tall 'Swedish Doll' grows, the blooms remain miniature, with the leaves and stems in proportion.

Between these extremes are hundreds of other varieties. Because of the range of heights, it is difficult to talk with any real meaning about an average height. Many miniature roses reach about 35 to 45 cm but their final height depends nearly as much upon the growing conditions as it does upon the variety. Plants which are frosted to ground level in the cold of a Boston winter are pruned back to 30 cm in Christchurch. It is no wonder plants in our Christchurch garden grow taller; they have a head start. In areas with warmer temperatures and more moisture, plants could well grow taller than ours. Only when you know something of the conditions under which the plant was grown does a given height mean very much. Whenever we mention the height of a plant, we will be talking about the way it grows in our garden where it has been grown with reasonable care and ample water. (See page 125 for a description of Christchurch weather and climate conditions.)

To the gardener, the way in which a bush grows is almost as important as its height. Some miniatures form dense, tight bushes with the flowers on short stems close to the bush. Others are more open, with a light, airy look, and the flowers carried on longer stems. True climbers will grow long shoots which need to be tied to a framework, and the following season produce their flowers on short stems along the length of the shoots. Another form of miniature rose creating a lot of interest grows long shoots like climbers but, instead of reaching upwards, these arch over and, 50 or 60 cm later, almost reach the ground again. These plants, such as 'Red Cascade' or 'Sugar Elf', are most attractive in flower as they produce their blooms along the length of these arching branches. The result can

be a mound of flowers 1.5 m across.

A few varieties with lighter stems lack the strength to arch but creep along the ground. One of the earliest of these was 'Nozomi'. In one year its growing shoots may reach out 1.5 to 1.8 m and in time cover an area 3 m across. Others, such as 'Snow Carpet', are less vigorous but provide a closer cover. These arching and trailing plants are ideal on a bank or in a container where the hanging branches can be seen to best advantage.

Height and growth vary, but flower form, colour and size vary even more. Many of the earlier varieties had flowers similar to the species roses to which they were closely related. Typically they were small, informal flowers which opened flat with a large number of narrow, short petals, usually in various shades of pink. Some of the more recent varieties, such as 'Stacey Sue' or 'Seabreeze', still have this form, but in every other way they are much improved. They have many more flowers which last longer and the plant is much more reliable. A flower which opens flat is found in many other miniatures, some of the better known are 'Fiesta Gold', 'Wee Man' and 'Little Buckaroo'. As opening blooms, they have attractive buds of classical form, but in a few days they have opened out. Even as fully open flowers they have their own special appeal, with their broad petals and bright colours. Although some have many petals, others have as few as ten or fifteen. For those who seek charm in simplicity, it is even possible to find examples of the true single rose. One such rose is 'Simplex', with its five broad, wavy petals of apricot changing to cream surrounding its prominent gold stamens.

When people think of roses, they more often picture the high, centred bloom with its attractively arranged reflexing petals. If someone tells us that they want a variety with true rose flowers, we know that this is what they mean. More and more of the modern varieties have this form; others have been around for several years. One of these, and still one of the best, is 'Starina.'

Much of the appeal of the early miniature roses must have been in their small flowers, often less than 2 cm across even when they were fully open. As other types of roses were crossed with miniatures to increase their strength and give greater variety of flower form and colour, the flower size increased. As a result, miniature rose flowers now range in size from the 1 cm blooms of 'Si' to the large 7 cm blooms of 'Angel Darling', 'Kaikoura' and the like. When they reach this size, the distinction between miniature roses and small floribundas becomes very blurred.

Remember those hybrid teas and floribundas? Miniature roses also come as smaller versions of these and all the intervening types. The first miniature roses often had only a few flowers on each short stem, but as the influence of the floribundas was felt, stems became longer with more flowers. Some even had true floribunda-like flower heads with many blooms. More recently, a number of varieties have appeared where blooms occur one to a stem. Often these are hybrid tea type flowers with longer stems which are ideal for picking.

There is, then, no such thing as a typical miniature rose. They vary so much in height, flower form, size and colour that it is impossible to point to any one plant and say, 'That is what we are talking about!' When we use words such as 'smaller' or 'higher', however, we must have some basis for comparison. So we will take our courage in both hands and describe our own concept of the average miniature rose. When we talk about such a rose, we have in mind a bushy plant which grows about 40 cm tall. (It may carry its flowers above this but dead flowers are always going to be removed.) Its flowers have fifteen to twenty-five petals and are about 2 to 3 cm across when fully open. While there are often only one or two open blooms on a stem at any one time, there are usually several buds to a stem. Its normal leaves, made up of five leaflets, are about 4 or 5 cm long, narrow and pointed, and have serrated edges. It has small, fine thorns, including some along the reverse side of the mid-ribs of the leaflets. This will be our miniature rose when we need a standard for comparison.

ROOTS

Oh, no man knows
Through what wild centuries
Roves back the rose!

Walter de la Mare — *All That's Past*

Somewhere in a book about miniature roses something must be said about roots — not the sort that hold the plant in the ground and feed it, but historical beginnings. If you find family trees and a bit of history tiresome and uninteresting, then leave this chapter and come back to it when you are more acquainted with miniature roses and you want to find out more about them.

When Walter de la Mare wrote the words which introduce this chapter, he was almost certainly not thinking of miniature roses, but he could have been, for the sentiment is just as applicable. Many think that miniature roses have been about for only a short time. Others are aware of a romantic story which tells of the discovery of the first miniature rose early this century. What are the true origins of our miniature roses?

Roses are among the most ancient of plants. Fossil forms of the rose have been found which pre-date man by many thousands of years, but only in the Northern Hemisphere; no native roses have been found south of the Equator. While a variety of rose species have been found growing wild in the temperate areas of Europe, Asia and North America, the only roses in New Zealand and Australia are those which have been taken there by man. Even those roses which now grow wild in some of our country districts have their origins in roses planted by the early European settlers.

References abound which show that roses were known and cultivated by the ancient civilisations of the Mediterranean. Selected forms were grown and a thriving trade in rose petals and rose oil existed in Greek and Roman times. Rose petals were frequently used to cover the floor on special occasions; imagine the number of roses used when Antony and Cleopatra walked knee deep through petals. There were, however, dangers; on at least one occasion an inebriated guest at a Roman banquet sank among the roses and suffocated.

Roses were also cultivated in the East where the Japanese and Chinese were accomplished gardeners. Chinese screen paintings more than a thousand years old show the distinctive flower and foliage of the China rose, *R. chinensis.*

Modern roses differ considerably from these ancient varieties. They are the result of the intermingling of the species and of careful breeding and selection over many years. The same is true of modern miniature roses.

In 1917, while climbing in the Swiss Alps, Dr Roulet, a major in the Swiss Army, came across a small rose growing in a pot on a cottage window ledge in the small village of Mauborget. This rose was tiny but there must have been something else which made it special. At an altitude of 1200 m the winters were hard. Although the villagers considered the rose delicate and unsuitable for outdoor growing, they had taken the care to grow it for more than a hundred years. In this time they had found that it grew best on a window ledge where it received the sun and light necessary for it to flower.

Roulet reported his find to a nurseryman friend, Henry Correvon, who was growing some small roses, but none quite like this. Correvon went to Roulet, intending that they should see these plants together, but before this could be done, the village of Mauborget was destroyed by fire. Fortunately, the same rose was growing in another nearby village and they were able to get some small pieces. Some of these cuttings grew and from them hundreds of plants were propagated. This was the beginning. From these small plants, or their offspring, have come all our modern miniature roses.

This little Swiss rose was named *R. rouletii.* At first it was thought to be a previously unknown species, but no plants have ever been found growing in the wild. Other miniature roses were known to exist — at least one of these, the 'Lawrance Rose', was being grown at Floraire by Correvon — but *R. rouletii* was different. If it didn't originate in Switzerland, where had it come from? It must have been a variety from some earlier time; a variety which had appealed to these villagers who had taken the trouble to preserve it for so long that nobody knew its origins.

Look up 'Rouletii' in three different books and it is possible to find three different spellings. Correvon named the rose after Dr Roulet but even he used two different spellings, both *R. rouletii* and *R. roulettii*. On occasions 'Rouletti' also appears. The same variation in spelling occurs with other old rose names. The 'Lawrance Rose' is variously spelt 'Lawrance' and 'Lawrence'.

When rose growers or hybridisers need information about the origins and parentage of a rose, they frequently turn to the book *Modern Roses*. In this are found descriptions of all roses registered over the last sixty years or so, together with any known details of earlier roses which are important in some way. In the case of *R. rouletii, Modern Roses* is not very helpful. It reads:

> Rouletii. Min. (Discovered growing in pots on window ledges of Swiss cottages by Major Roulet; int. Correvon.) Very small (less than ½ in.), dbl., rose-pink. Dwarf (6 in.); long blooming season. Very hardy. Should be grown in poor soil to keep it dwarf. See *R. chinensis minima*.

Against *R. chinensis minima, Modern Roses 8* gives a number of alternative names. They include 'Rouletii', a description and the date, 1815. It is to the early 1800s we must go if we are to find the origins of the modern miniature rose.

Before 1800 most roses grown in European gardens were descendants of the French rose, *R. gallica*, or closely related to it. The different varieties were largely the result of mutations ('sports') or natural hybrids, where different varieties planted close together produced chance seedlings. Most of these early roses flowered only once a year. It was not until the later part of the eighteenth century, when artificial or hand pollination was first used, that some control could be exercised over the choice of parents.

This latter half of the 1700s was a time of exploration and discovery. Ships from Europe were venturing to new areas of the globe, and with the increased interest in science, in them sailed the botanists. When Captain James Cook came to the Pacific and New Zealand he had with him the botanist, Joseph Banks, whose task was to describe and collect previously unknown plant material and bring specimens back to London's Kew Gardens. This was the time when the great trading companies were setting up their stations on the other side of the world. Alongside their depots and residences were their gardens where it was possible to grow new, exciting and exotic plants which would find their way back to England and Europe. It is no wonder that at this time there was an increasing interest in the flower garden and a tremendous influx of new plant material into Europe.

At about this time, new rose introductions to Europe and England included, perhaps for the first time, plants originating from *R. chinensis*. *R. chinensis*, the 'China Rose', is a large, straggling, vigorous climber which can grow 6 m across. It has single pink to crimson flowers about 5 cm across, usually several together, on long stems. It was not these large species which were brought into England, but smaller cultivated varieties. They were to have a tremendous effect upon the development of roses as they were recurrent flowering.

We do not know whether these smaller forms arose naturally as mutations of *R. chinensis* or whether they developed through the persistence and skill of oriental gardeners, who over many years sowed seed selected from only the smallest plants. The ancient art of bonsai has long existed in the East and it is quite possible that gardeners would have searched for and cultivated plants more suited to the garden, and which could also be fashioned into suitable bonsai specimens. Collected Chinese paintings show sixteen different roses known to be growing in Chinese gardens at the beginning of the nineteenth century. Tess Allen describes them in the *National Rose Society Annual* of 1973: 'There are double, single, large, small and miniature roses; some borne singly and others in clusters; the petals of one rose have scalloped edges. The colours are white, purple, violet, pink, red, buff-yellow and golden-yellow.'

Among the early introductions was 'Slater's Crimson China', brought into England by Gilbert Slater, a director of the East India Company, in about 1792. This rose was also known as *R. semperflorens*, the 'Ever-Blooming Rose'. Even when grown as a greenhouse plant, it reached less than 90 cm.

Another even smaller rose was the 'Dwarf Pink China'. While some say that this was a sport, or perhaps a seedling, found in England, it is generally accepted that it is of Chinese origin and came to Europe by way of Singapore, India and Mauritius, arriving in England about 1805. Perhaps it was the double, purple-red miniature rose illustrated in the Chinese paintings already mentioned. Whatever the source, by 1829 at least eight different miniature variations of this rose were in cultivation.

These plants which are thought to be descendants of *R. chinensis minima*, or perhaps even variations of this, were known by a variety of names. One of these, 'Fairy Rose', is still sometimes used today. They were also known as 'Miss Lawrance's Rose'. A popular painter of the day, Miss Mary Lawrance, was often asked to paint 'portraits' of new varieties of plants to popularise them as they became available. In 1799 she had published one of the first books devoted entirely to roses, *The Collection of Roses from Nature*, which included ninety of her paintings. Miniature roses were not included as they had not yet arrived in England, but it is possible that she painted them later. Whatever her association with miniatures, one prominent nurseryman, no doubt with an eye for publicity, called his miniature rose 'Miss Lawrance's Rose' or *R. lawranceana*.

An illustration of one of these roses does exist, in Curtis' *Botanical Magazine* for 1815. Plate 1762 is coloured and shows a branched stem with two flowers and a bud. The flowers are single, with only five petals which are white to pale pink with a deeper pink shading on their upper margins. It could be a painting of any one of hundreds of seedlings which are likely to occur and be discarded when seed from miniature roses is planted. It is the comments with the illustration which are more interesting. The drawing itself is labelled *R. lawrenceana* while the Latin description is headed *Rosa Semperflorens Minima*, and 'Miss Lawrance's Rose'. The actual comments are worth quoting in full:

Several varieties of the Rosa semperflorens, differing in size, colour, and scent, have, within these few years, found their way into the different collections about town, and have generally been represented as fresh importations from China; we believe, however, that most of them have been raised from seed here. Every experienced cultivator knows, that the varieties to be obtained in this way are endless.

Our present subject is the most dwarf Rose that has ever fallen under our notice, rarely producing any branches, so large as represented in our plate. We are inclined to consider it as a mere seminal variety, perhaps of hybrid origin; yet we cannot assert that it is not a distinct species. It is generally known among collectors by the name of MISS LAWRENCE'S ROSE.

The plant from which our drawing was taken, was communicated by MR HUDSON, of the war-office. Flowers most part of the spring, and has an agreeable, though not powerful scent.

It is clear from these comments that several variations of the illustrated rose existed and that these were probably seedlings from some original plant. These seedlings varied considerably in size and form because when a rose is raised from seed 'the varieties to be obtained in this way are endless'. Undoubtedly selected seedlings would be kept and propagated and *R. lawranceana* became not one but a class of roses. Some of these were given names, frequently descriptive words such as 'Alba', 'Blush' or 'Pigmy'. None of these varieties can be identified as roses grown today.

Miniature roses were also painted by the greatest rose painter of all time, Redouté. At about the same time that the China roses were coming to England and Europe was engaged in fighting the Napoleonic Wars, the Empress Josephine was establishing her residence and garden at Malmaison. The garden contained a wide variety of plants, but is best remembered for its roses; to it were brought all the known varieties in Europe. Ships captured at sea were searched for new plants, and a botanist went with Napoleon to Egypt in search of new material for Malmaison. Even though England and France were at war, Josephine remained in contact with the nurserymen of London and the shipment of roses continued through the blockade of the French ports. Just as fashionable women of this time sat to have their portraits painted, so Josephine had her flowers painted. Born in Belgium in July 1759, Pierre-Joseph Redouté had already had volumes of flower paintings published before he was commissioned to paint the roses of Malmaison. Josephine did not live to see the task completed as she died in 1814, three years before the paintings began to appear. The publication of the three volumes containing 169 plates was not completed until 1824.

Whether miniature roses were actually grown at Malmaison is not known, as no accurate list of all the roses has ever been found, but Redouté did paint them. There are two paintings. One is entitled *Rosa Indica Pumila, Flore simplici*, the other *Rosa Indica Pumila, Flore multiplici*. Both paintings show much branched plants with long, pointed leaflets. The buds and flowers are carried on long stems and new growth appears close to the spent bloom. All this is typical of our modern miniature roses. One of the paintings shows a simple, single bloom with five brick red petals — *Flore simplici*. Double blooms with many more petals appear in the second painting. *R. indica pumila* is now believed to be *R. chinensis minima*.

Whatever the origins of these little roses, they seem to have become very popular between 1830 and 1850 when more than twenty varieties were included in rose lists of the time. They were certainly sold in England and France. The double form had been sent from England to France where it was known as 'Bengal Pompom'. Was this the

Rosa Indica Pumila of Redouté's painting?

There were other miniature type roses. Some, like *R. banksiae* or *R. multiflora*, do not really belong, as their clusters of miniature flowers are grown on monstrous plants. Others, though, did grow small flowers on small bushes. We have recently obtained a plant of 'Pompom de Bourgogne', a small rose belonging to the centifolia group and possibly another included in Redouté's paintings. Because of our interest in miniature roses, it has been exciting to see it growing and flowering. Only 40 cm tall, it shows a number of rose pink pompom flowers 2 or 3 cm across.

One of the earlier books on roses as garden plants was written in 1848 by William Paul of Cheshunt, England. In *The Rose Garden* he wrote of some of these other miniature roses:

THE MINIATURE PROVENCE, OR POMPOM ROSE
The Roses in this group are remarkable for their diminutiveness. They are well adapted for edgings to the Rosarium, or as Rose-clumps generally. They are sometimes planted in masses, in which manner they look well, as they are of neat growth, and bloom profusely; but they do not last long in flower: and for this reason we should hesitate to recommend them, except under particular circumstances. The Chinese and Bourbon Roses are usually preferred for this purpose; and no wonder, when it is considered that they produce their beautiful flowers during one half of the year.

This was the secret of the popularity of the new breed of miniature roses: they bloomed not only in the spring, but continued to bloom through the summer months. William Paul also wrote about these roses:

THE LAWRENCEANA, OR FAIRY ROSE
The first of these interesting Roses was introduced from China in 1810. The varieties form pretty objects cultivated in pots, rarely exceeding a foot in height. Thousands of them are sold in our markets every year, and beautiful they are when covered with their tiny blossoms.

We sometimes forget that the 1800s was also a time of great expansion and settlement in the United States. Robert Buist, a nurseryman and florist of Philadelphia, also knew and grew these new miniature roses and in *The Rose Manual* of 1844 he writes: 'these roses have been the produce of seed saved from the smallest flowers of the kind, year after year, till they now become the fairies of the tribe.' He describes several varieties, including 'Master Burke', and he quotes the following from another publication:

When three years old, the Master Burke had fine full bloom and very double flowers; and the half of a common hen's egg-shell would have covered the whole bush without touching it. This I saw and assert to be a fact. It is now seven or eight years old, flowers regularly every year, affording wood for propagation, and has never attained two inches in height, nor its whole top exceeded one, or one and a half inch in diameter; the rose is about the size of a buckshot.

We may have doubts about a rose of this size but he assures us that 'the article was written by a gentleman of high standing'.

The roses were not cheap. In 1838 C. M. McIntosh listed about twenty varieties and in many cases included prices (in shillings and pence). Some were:

Blush, 1/-
Minima, bright rose, 1/6
Nigra, very dark crimson, 2/6
Pallida or Alba, rose-tinged, 2/6

Remember that this was in 1838 when a pound of coffee cost 16 pence and the village schoolmaster was being paid £75 a year. Compare these prices with roses being sold in New Zealand fifty years later when bush roses were advertised: 'The best varieties in this list, one shilling each or nine shillings per dozen. Except where otherwise priced good old varieties ten pence each, or eight shillings per dozen.'

It seems that miniature roses, like many other plants at this time, could be afforded only by the wealthy or the enthusiast; they were novelties and collector's items. They were frequently grown as pot plants and it is easy to imagine them standing in the conservatory or being brought, in flower, into the Victorian drawing room as a curiosity.

And, like so many curiosities, they had their place for a time and were then discarded. By the end of the century they had all but disappeared, and there are a number of reasons why this may have occurred. It may have been due to the way they were grown. Miniature roses can be grown successfully in a pot, but not without the appropriate conditions of soil, light and humidity. Consider the number of plants, such as begonias and cyclamen, which are bought in flower today, grown in a pot for a few months (or even years) but are thrown away when the flowers have gone, interest wanes and the plant deteriorates. How many miniature roses flowered in their pots for a few weeks and were then consigned to the rubbish heap?

Perhaps it was more a matter of fashion; there are changing fashions in plants. What has happened to the aspidistra, that plant which did so well in the dim light of the Victorian hallway? It was not hard to grow but is no longer fashionable. Was it a case of the rose of the fifties being replaced by something else in the eighties? This was the time when the tea rose was at the height of its popularity, soon to be replaced, in turn, by the hybrid tea. 'La France', often considered the first hybrid tea rose, was introduced by Guillot in 1867. 'La France' grew like a hybrid perpetual, but it had the fine shaped buds and free-flowering character of the tea roses. It was the beginning of a new trend in roses, and another new type of rose was soon to appear.

During the 1860s seed of the wild species *R. multiflora* (sometimes known as *R. polyantha*) was sent from Japan to France. In his nursery at Lyon, Guillot planted seed from a cross between *R. multiflora* and the 'Dwarf Pink China'. It would be hard to imagine greater extremes of plants. *R. multiflora* is a vigorous, large shrub or climber which each year grows arching stems up to 3 m long. As its name suggests, it has masses of flowers in large branching clusters up to 30 cm across. Individual flowers are single, white and usually less than 2.5 cm, with the stamens making a prominent golden centre. It is used extensively in New Zealand as the understock for bush roses and you may have had some in your garden when a sucker appeared on one of your roses. The 'Dwarf Pink China' is, of course, *R. chinensis minima.*

From these two diverse parents, the seedlings showed considerable variation. The first generation seedlings were mostly double flowered but they were still climbers and once flowering like *R. multiflora.* From these first plants a second generation contained some seedlings which were dwarf and repeat flowering. They included both 'Paquerette', a white, and the pink-flowered 'Mignonette'.

These were the first of the dwarf polyantha roses, combining some of the better characteristics of both parents. Dwarf and with a prolonged flowering period, they were, at the same time, hardy and the flowers were borne in large clusters, making a good display. These roses were introduced in 1875 and 1880 and it would not be surprising if, over a number of years, they replaced the early miniature roses as the favourites. Although they were still small, they were stronger growing and had a much greater range of colours. Some of these dwarf polyantha roses appear in our story again later, so we will stay with them a little longer.

From these original plants came 'Gloire des Polyantha' (1887) and its seedling 'Mme Norbert Levavasseur' (1903), which in turn was used as a parent of the first hybrid polyantha (later to become the floribundas). There was something unstable in these dwarf polyantha roses. Ann Wylie reported 262 varieties of dwarf polyantha in existence in 1954. Of the 150 whose origin was known, 81 or 54 per cent were sports or mutations and of these 29 were direct sports of 'Orleans Rose'. Another to produce several sports was 'Superb'. One of these was 'Gloria Mundi', the first known rose to contain the pigment which gives the

brilliant orange-scarlet hues of a number of our modern roses. When we see those brilliant orange-red colours of 'Super Star', 'Alexander' and others, it is 'Gloria Mundi' and its sport 'Gloire du Midi' that we have to thank.

Another branch of the polyantha family gave us 'Cécile Brunner' (1881), often called 'The Sweetheart Rose', and 'Perle d'Or' (1884). These have small flowers but are not really miniatures. They are both still commonly grown.

By the end of the nineteenth century, New Zealand had a number of nurserymen propagating and selling roses. Although many lists included polyantha roses, we have been unable to find reference to *R. chinensis minima* or *Lawrenceana* varieties. Typical of the listings is this, from the 1899-1900 catalogue of D. Hay and Son, Parnell, Auckland:

Polyantha Roses
Also called Miniature, or Fairy Roses.
The flowers are very small, and produced in profusion. The plants are of very dwarf habit, and continuous bloomers. The buds are most useful for button-hole and other bouquets.

Under this heading are listed a dozen varieties with descriptions. None belong with our miniature roses. Pity, too, the buyer who failed to read the individual descriptions carefully. Although the roses are described as 'of very dwarf habit', included are five most vigorous ramblers such as 'Crimson Rambler', obviously very highly regarded but definitely not miniature! ('Every garden in New Zealand should have one — extremely vigorous growth, making shoots 8 to 10 feet in a season — it cannot be surpassed. An altogether unique and charming rose.')

So, for a time, miniature roses, as we know them, dropped out of sight. They had not, however, been completely forgotten. Correvon, and undoubtedly others, grew some varieties. The evidence is there if we look for it. *The Standard Cyclopedia of Horticulture* (1917), includes *R. chinensis minima*, and it helps to bridge the gap in its entry, which reads:

Rosa chinensis var. minima.
R. Lawrenciana R. indica var. pumila
Dwarf shrub, usually not over 1 ft high, with small rose-red flowers about 1½ in. across; petals often pointed. There are single- and double-flowered forms. The Fairy Roses belong to this variety.

And so we come again to Roulet and his little rose growing on a Swiss window ledge. This time the path back seems much clearer, but we can never really be sure. Some forms of the little rose which was brought to England in the early 1800s were sold in France and from there they reached Switzerland, to reappear many years later. It seems reasonable, but who knows?

Exactly which variety became 'Rouletii' is never likely to be known. Was it the original double form of *R. chinensis minima*? Perhaps it was 'Pompom de Paris', which was selling in the street markets of Paris in the 1850s? It could be some unknown variety which was never really recorded 150 years ago. Perhaps the discovery of 'Rouletii' made people more aware of miniature roses for, in the early 1930s, another variety was found in Oakington in England. Although it has the same type of growth, it is not 'Rouletii', since it is described as ruby-crimson with a white eye. This variety, 'Oakington Ruby', is definitely another descendant of *R. chinensis minima*.

We grow 'Rouletii'; in fact we grow two 'Rouletii' from two different sources. Both of them have the same sort of growth and both fit the general description of 'Rouletii'. One seems to be a little more vigorous and the colour is a little richer than the purplish pink of the other. Which is correct? We don't know. Both could be. With many miniatures, we see more variation than this between plants of the same variety and at times between flowers on the same plant. It doesn't matter to us, but it is interesting to look at our plants and know that their history takes us back all those years.

CREATING NEW VARIETIES

It is not very often that new varieties of roses just happen. Most are the result of deliberate breeding and careful selection but there are many occasions throughout the history of the rose when an unusual flower or growth form has drawn attention to a new variety growing among established plants.

Sometimes these are mutations or sports. From time to time a change takes place in the growing shoot on a plant, and as the shoot continues to grow it develops some new characteristics which show that it differs from the rest of the plant. It may be the shape of the leaves; the flower may have more or fewer petals; the shoot may continue to grow and grow and become a climbing branch. Among our garden roses there are many climbing sports where both bush and climbing forms have identical flowers. One of them is identified by the word 'climbing' and so there is 'Climbing Peace' and 'Climbing Sutters Gold'. Climbing sports of miniatures are also known and in the United States varieties such as 'Climbing Mary Marshall' and 'Climbing Baby Darling' are grown. Occasionally the change works the other way and a vigorous climber or shrub produces a dwarf shoot. This possibly happened from time to time to give small forms of *R. chinensis*. It certainly happened with 'The Fairy', a rose no bigger than many of our modern miniatures. It is a sport from 'Lady Godiva', a vigorous rambling climber.

More often, or at least, more obviously, a change takes place in the colour. It may only be a slight variation or it may be that the new colour is completely different. We have at the moment a sport of 'Mary Marshall' which was brought to us by an observant rose grower in the North Island. In it the pink and orange of 'Mary Marshall' have become a yellow with a touch of apricot and red. (See page 90.) There are many questions which must be answered before it may be released as a new variety. Is it stable or will it revert and next season produce normal 'Mary Marshall' flowers? Will it propagate to give yellow or pink flowers? Was the colour the only change? Perhaps there are also changes in flower shape, vigour or disease resistance which make this new plant less desirable and not worth commercial propagation? At the moment this sport looks quite promising but it is early days yet. It is likely that more sports of roses occur than we realise, but many are insignificant or unnoticed and they are lost. Perhaps you have a sport in your garden?

Chance seedlings were another source of new varieties. Where mixed varieties were growing close together, pollen was carried by bees or perhaps even by wind from one flower to another to produce seed which fell into the ground and from it came a plant clearly different from its parents. Undoubtedly those early gardeners and nurserymen helped by collecting rose hips and, having planted the seed, they would watch with interest as the seedlings grew and flowered.

It is only over about the last hundred years that cross-pollination by hand has taken place. Techniques differ in detail but the essentials remain the same. Modern hybridisers frequently work under cover in some form of greenhouse. In this way they avoid delays and problems caused by rain, have earlier and longer seasons in which to work and lessen the danger of stray pollen from wandering bees.

Once the parent plants to be used have been decided upon, the first step in cross-pollination is to collect the pollen. This is done by removing the pollen-bearing anthers from the newly opened flower and storing them until the pollen is required. As they ripen the anthers shed their pollen which remains usable for some time.

Flowers on the other parent, the seed parent or mother plant, must also be prepared. Before the flower has opened, the petals are removed to expose the stamens and stigma. The anthers are then carefully cut off with fine scissors to prevent unwanted pollen falling on the stigma. The required pollen is brushed across the exposed stigma and the operation is almost complete. Some use a fine camel hair brush to do this but Sam McGredy finds that a forefinger moistened on the tongue does the job just as well! Finally a tag is tied to the flower stem to record the cross which has just been made. Over the next few weeks, as the seeds develop, the hip swells and later begins to change from green to orange and red. At this stage the hip with its label is removed from the plant.

If the ripe hip was left on the plant, sometime during the winter it would fall into the ground where the outside covering would gradually rot away to expose the seed. This seed would remain dormant through the heat of the next summer; the cold of another winter and the warmer days of spring are needed to start the seed into growth. The hybridiser has to match these conditions but telescopes them into a few months. Once the seed is removed from the hips, which have been stored in a cool place, it is usually placed in a refrigerator for several weeks, and is then planted in trays in a warm greenhouse. The seed thinks it has been through the second winter and begins to grow!

Most seedlings flower for the first time while they are still only a few centimetres tall. This is a time of great excitement. It is not always realised that roses grown from seed often have no resemblance to either parent. It is, in fact, the hybridiser's art to bring together the desired characteristics of each parent. Gradually the health of one is combined with the colour of another and later the form of a third until finally the perfect rose is found. The perfect rose? We haven't yet achieved this. There are some very good ones but they can be, and are being, improved.

It is an interesting experience to walk with Ralph Moore through his Californian nursery and seedling houses. As you walk, the conversation goes rather like this: 'Wouldn't it be interesting to put this flower on this plant and we could change the colour. . . . ' And it is people like Ralph who have the skill and experience to do this. Years of observation have shown him which plants to use to introduce the required characteristics. But, even with all the knowledge and planning, there is still the excitement of the seedling flowering for the first time. There is nothing Ralph enjoys more than having another miniature rose enthusiast with whom to compare impressions of the new seedlings and share the discovery of that exciting new variety. On that last morning of our stay, with the suitcases in the car and a plane

to catch in Los Angeles, we still made time for a last look when Ralph said, 'Come and have a look through the seedlings to see what's showing this morning!'

The excitement is there because each plant, as it flowers, is unique; even with all the knowledge available, no grower can be sure about what will occur. The result may be close to what has been aimed for, or it may be nothing but rubbish. Sometimes the unexpected happens and in the batch of seedlings a winner appears which is completely different from what was expected. The chance of a real winner is, however, very small. We were walking recently with Sam McGredy through his hybridising house in Auckland, New Zealand. He told us that in the current season he hoped to make 7 000 crosses from which he expected to get 5 000 heps. These should give him about 50 000 seedlings and from them he may get one real winner. He, of course, is aiming for hybrid tea and floribunda roses as well as miniatures, but the odds remain the same for the miniature specialist.

You may not be a Ralph Moore or a Sam McGredy but you can share some of the excitement of producing your own miniature seedlings. Look over your miniature roses this autumn and, if there are any heps, remove them. Cut them open with a sharp knife and take out the seed. (There may not be many seeds in the hep, as miniature heps frequently contain only one or two seeds.) Wrap the seeds in damp paper (a paper towel is ideal), put that in a plastic bag and place the whole package in the refrigerator for about six weeks. At that stage, plant the seed a few centimetres deep in a tray of potting mix and place it in a glasshouse or in the shelter of a fence or building. Within a few weeks the first plants will begin to show. Only a small fraction of the seeds are likely to grow, but whatever happens and however poor or good they are, you will experience the thrill of seeing your first rose seedlings. Remember, they are unique — and they are yours!

The first flower gives some indication of the plant's potential. By that time there will also be signs of disease resistance or lack of it. Any plants which are prone to disease should be removed. The colour of the flower will not change but the next flower will be larger and may have more petals. We asked Sam when he decided he had a miniature. He smiled and replied, 'I don't worry. As long as it is healthy and a bushy grower with a good flower, it goes outside.'

Seedlings which show promise are grown on and other plants are propagated from them. These are watched very closely over the next few years as they grow. Any which do not meet the grower's standard are discarded. Perhaps the flower is too large or the form of the bloom is not good enough. Perhaps it is just too similar to some other variety. Whatever the reason, only the very best survive to be put on the market. And, of course, before that new variety can be sold that one plant must be increased to thousands. All that takes time, as Sam McGredy knows. We were discussing a new variety he was introducing into his breeding programme, and Sam commented, 'If anything good comes from them, it will be at least ten years from now before it gets on the market.'

Before a miniature rose is sold, it must be given a name, and the name is important. Back in the 1950s, when miniature roses were considered of special interest to children, names associated with nursery rhymes and stories were common and, among others, we have 'Bo Peep' and 'Cinderella'. Many names honour a person. 'Mary Marshall' is named for a charming rose grower from San Mateo, just out of San Francisco.

For many years she and her husband, Don, have devoted many hours to roses and the American Rose Society. They clearly display their enthusiasm. In California, where you can have personalised car number plates, Don and Mary's say 'ROSENUT'! And now there is a dark red miniature rose called 'Don Marshall'. Sometimes plants are named after members of the family or family friends. 'Debbie' is a grand-daughter of Ralph Moore, as is 'Sheri Anne'; 'Baby Katie' is the first grand-daughter of American grower Harmon Saville.

Some names relate to events or places. What better name could there be for a red and white striped rose which came out during the American bicentennial year than 'Stars 'n Stripes'? Sam McGredy has a whole series of miniature roses with New Zealand names such as 'Wanaka', 'Kaikoura' and 'Moana'. Names such as 'Red Cascade' and 'Snow Carpet' vividly describe the rose in question and even names such as 'Yellow Doll' and 'White Angel' tell us something about the varieties they identify. Other names have an appeal of their own and just seem to fit. What better can there be than 'Cuddles' or 'Party Girl'? These names aren't easy to find and a number of raisers keep notebooks in which they record possible names as they are suggested by friends or circumstances.

If a rose is to be sold, the name is usually registered. This means that the same name cannot be used by somebody else for a different rose. This is an international arrangement and the American Rose Society acts as the clearing house for rose registrations, each year publishing a list of new varieties in its annual. Modern roses are frequently given a variety code name by the raiser. Frequently this is related in some way to the raiser's name. McGredy-raised varieties usually begin with 'Mac', as in 'Macbipi', while Ralph Moore uses 'Mor' in code names like 'Morgal'. These names stay with the variety no matter where it is sold. This means that although a rose may be sold with different names in New Zealand and the United States, it is still possible to keep track of it through the code name. In the past, different names for the same rose could cause problems. Some years ago we were importing some new miniatures from England. There was an orange-yellow variety called 'Darling Flame' which sounded very attractive. We ordered it and it was not until after all the complications of permits and quarantine that we discovered, as it grew and flowered, that we had grown the same rose for several years under the name 'Minuetto'. It sometimes makes good sense to use different names in different countries. How many people in New Zealand would buy a rose called 'Zwergkönig'? They might buy and grow the same rose as 'Dwarfking'.

This is probably an appropriate place to say something about plant patents or protection. The breeding and introduction of a new rose involves raisers in considerable expense, so it is only right that they should receive a reasonable return for their work. Previously, once a new variety had been placed on the market, other growers were able to obtain plants from which hundreds of others could be propagated and sold without any return to the raiser. Now in many countries, including New Zealand, it is possible to obtain a plant patent or protection order. This means that once a variety has had a protection order placed on it, only licensed growers, who pay a royalty for the privilege, are allowed to propagate and sell that variety. This does not mean you cannot grow a few plants for your own use, but growing for sale or any commercial purpose without permission is illegal.

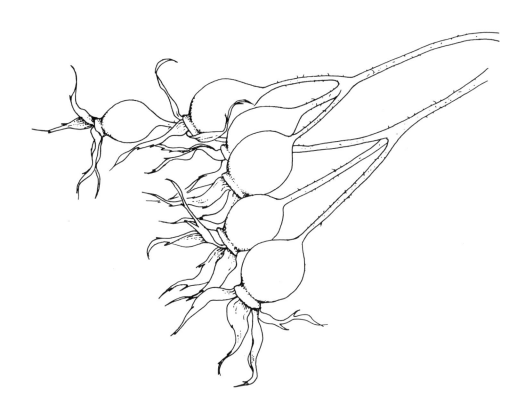

TWENTIETH-CENTURY MINIATURES

Even though 'Rouletii' was rediscovered in 1917, it was not until the 1930s that other miniature roses were known. 'Rouletii' was grown and propagated by Henry Correvon and released in 1922; other nurserymen also grew plants and once again miniature roses appeared in the Paris flower markets. 'Rouletii' had a wide distribution throughout Europe and the United States. In England, where it received a Medal of Honour, it is reported to have been very popular. From Germany, came the statement that it was exactly the same as 'Pompom de Paris' which was already in cultivation there.

In 1933 while Robert Pyle of the Conard-Pyle Co. (Star Roses) was in Holland, he called upon Jan de Vink and found him 'playing around with the breeding of miniature varieties, just for his own fun'. De Vink had been using mixed pollen from whatever happened to be available with 'Rouletii' as seed parent and he had a few seedlings from a batch where he had used pollen from 'F. J. Grootendorst', 'Gloria Mundi', an unnamed yellow hybrid tea, and a seedling from 'Tausendschön'. Perhaps time adds romantic details, but a recent article in the 1983 *American Rose Annual* says:

Thirty seeds were produced, but only fifteen germinated. The seedlings were another rather sad story. They were weak and gradually died off until only four remained. However, three showed encouraging dwarf characteristics. Flower buds were forming, but at the same time the plants were failing rapidly. De Vink admitted his situation was desperate, so he gambled on emergency surgery, snipping the plants just above the seed leaves creating cuttings. Two of the cuttings succeeded. One bore "Peon", so named, de Vink said, because "it seemed like a pawn on the chessboard". The white-centerd dark red dandy was subsequently delivered to Robert Pyle [who] renamed it "Tom Thumb".

It is 'Gloria Mundi', an orange-red polyantha, that is accepted as the probable pollen parent of 'Peon', since it is most likely to have produced a true miniature with red blooms even smaller than those of the pink 'Rouletii'. It is interesting to realise that 'Peon' was a direct descendant from *R. chinensis minima*, not only on its mother's side, but also just a few generations back on its father's side, as 'Gloria Mundi' also has *R. chinensis minima* in its parentage.

In the United States 'Tom Thumb' became the first miniature rose to receive a plant patent. No doubt encouraged by the royalties he was receiving, de Vink continued to experiment with miniature roses, but it was not until 1938 that he sent his second to the United States. This was 'Pixie', which was released in 1940. At about this time, the war in Europe interrupted the production of further miniature roses, but other varieties such as 'Midget', 'Sweet Fairy', 'Bo Peep' and 'Red Imp' were to come later. One of his last, 'Cinderella', is still regarded as one of the best of the really small white varieties.

To end this description of the part played by de Vink in the development of miniature roses let us quote from 'Modern Miniature Roses', an article in the *American Rose Annual* of 1953, by Roberta Lord of the Conard-Pyle Co.

The most astonishing part of Mr de Vink's success is the number of fine varieties he has produced out of what is really a very small-scale operation, in more ways than one. He raises only a small number of seedlings a year, in his greenhouse or in coldframes in the backyard behind his little Dutch house. His whole nursery is less than an acre in size and mostly devoted to the growing of small evergreens. This nursery, with its single greenhouse that is probably not more than 13 by 30 feet overall, is about half a mile from his home. He bicycles back and forth by a footpath and a little footbridge over a canal. To lock up his nursery for the night, he raises the footbridge! The whole establishment is pixie-like, but it has become a big success in the world of roses.

Another source of miniature roses in the 1940s

was Pedro Dot of Spain. Like de Vink's, his first miniature roses, 'Perla de Alcanada' ('Baby Crimson') and 'Perla de Montserrat', were limited to pinks and crimsons. The first yellow miniature, 'Baby Gold Star', was introduced in 1940, but it was not until 1951 that the real breakthrough in colour came with 'Rosina' (also known as 'Josephine Wheatcroft'), a golden yellow still popular today.

Pedro Dot died in November 1976 in the small town of St Feuil de Llobrigat, near Barcelona, where he had been born ninety-one years earlier. In the *Rose Annual* of 1977, Harry Wheatcroft wrote:

I first met Pedro Dot in the 1930s and fell in love with his miniatures which had been of little interest up until then. He offered me a selection to test and by the time we had built up a stock the war came and we ceased growing roses until the war was over. When we were able to release them they aroused great interest — varieties like "Perla de Alcanada", "Perla de Montserrat", "Rosina" and others were eagerly sought after and soon hybridists throughout the world were following the path that Pedro Dot was treading — until now, there is a large demand.

While some of Dot's later varieties, such as 'Pilar Dot', have larger flowers and grow more strongly than true miniatures, he also raised 'Si', the smallest of all miniature roses.

Although others raised miniature roses in the years following World War II, it was Ralph Moore who led in the development of modern miniature roses. As he tells in his own small book, *The Story of Moore Miniature Roses*, Ralph Moore had an early interest in plants and roses and as a boy in the 1920s was growing seedlings from 'Cécile Brunner'. He saw his first miniature rose, *R. rouletii*, in 1935, and soon after obtained plants of this, 'Tom Thumb' and 'Oakington Ruby'. It was 'Tom Thumb' crossed with 'Carolyn Dean' which gave the seedling 'Zee', a major breakthrough in the breeding of miniature roses. That was 1940. 'Carolyn Dean' was a rambler raised

by Ralph Moore which, although it had only five petals, had long, attractive buds which, through 'Zee', it passed on to its offspring. 'Judy Fischer', 'Beauty Secret', 'Mary Marshall', 'Yellow Doll' and many others all have 'Zee' in their background. It is this 1960s generation of miniature roses which gave a range of plants with hybrid tea type flowers so different from the smaller, short-petalled flowers which had been more common among the earlier varieties.

Like all plant breeders, Ralph Moore developed his own breeding lines and, from his observation of numerous crosses, it was apparent that only one parent had to be a miniature, as the miniature factor seemed to be dominant. It may have been an inhibiting influence which limited growth and was related to fertility, as many of the early miniatures were sterile, or inclined to be. Very few varieties would produce viable seed, but fortunately several gave usable pollen which was used on floribundas or other varieties producing good heps. One popular seed parent was 'Little Darling', a floribunda which also bequeathed to its offspring its attractive flower form. As well as 'Zee', other seedlings which have never been released have played an important part in the breeding of Moore miniature roses. Unlike some growers, Ralph Moore has made no secret of the parentage of his miniatures. As he has said, 'I don't mind giving you the recipes when you don't have all the ingredients.' But these miniature roses appear in the background of most modern miniatures bred by other growers.

Ralph continues to raise miniature roses. He has raised more successful varieties than any other grower and, until a few years ago, probably more than all the others combined. His recent varieties are still among the best. His recent yellow miniature, 'Rise 'n Shine', is regarded as the standard against which all future yellows must be judged, and his 'Sierra Sunrise' was the best miniature for 1984 in the New Zealand trial grounds. (As well as the observations and tests of a new variety by the raiser, many countries have trial grounds where a rose can be grown and watched and compared with others for two or three years. The New Zealand trial gardens, controlled by the National Rose Society, are in Palmerston North. All types of modern roses are included and, although the plants are identified only by a number, the gardens are worth a visit by the keen rose grower.)

Ralph Moore has not stopped there. When we first met him back in 1971 we walked around our garden and talked roses. While among the old roses, he talked of his hope to duplicate completely a rose garden in miniature. We looked at the old striped rose, 'Rosa Mundi', and he spoke of his plans for striped roses. He talked, too, of his new moss roses and the many years it had taken to breed moss onto miniature roses. (Since that time a number of miniature moss roses have come to New Zealand. These are described later in this book.) In 1976 the United States celebrated its bicentennial year and it was in this year that that first striped miniature rose, 'Stars 'n Stripes', was introduced. While we were at his nursery in Visalia, California, in 1979, it took only a few minutes for Ralph to collect a dozen striped flowers from his seedlings.

There have also been the spreading varieties like 'Red Cascade'. Apart from their role in the garden, Ralph sees an important place for this type as plants for containers and baskets. And so the development goes on. And as we walked and talked, Ralph spoke of his dreams: 'Wouldn't it be great to have a cascading miniature which grew well in a basket, which was continually in bloom and also had striped flowers?' We wouldn't be at all surprised if it is not already growing in his greenhouse.

Today there are many rose hybridisers throughout the world who include miniature roses in their breeding programmes; most of the major hybridisers have at least one or two miniatures among their introductions. The Dicksons and the Meillands, the Kordes and McGredys, all have miniature roses to their credit.

In his hybridising house in Auckland, Sam McGredy has a number of miniature roses which he is sprinkling with pollen in the hope of something new. His main aim? Novelty — something different, perhaps a different colour, perhaps a stripe. It may be a different form of growth or shape of flower, but above all it must

be healthy, grow well and be almost constantly in bloom. The varieties he has released show these features. 'Snow Carpet', while it is not always in flower, is unique. It crawls along the ground and forms a dense mat. 'Wanaka' has an unusual flower form and is one of the most brilliant orange-red miniatures you will find. One of his most recent varieties, 'Ragtime', is the first 'hand-painted' miniature rose. Many of his varieties are 'patio roses', larger than those typically thought of as miniatures. The flowers are larger and the plants more vigorous but they repeat quickly and cover themselves with bloom.

This tendency reflects the European market where the demand is for larger, healthy plants with bushy growth which can be used for mass planting. To produce a show of colour, more prominent flowers and a mass of bloom are needed. As Wilhelm Kordes told us, 'In our company we are breeding mostly for more vigorous miniatures, as this aspect is the most important for a good sale. We are also looking for bigger flowers.'

These are, of course, dwarf floribundas, the patio roses mentioned earlier. In a letter, Patrick Dickson of Dickson Nurseries in Northern Ireland writes, 'Though we are not involved in breeding miniatures as such, we do use some in our programme producing what we call "Patio Roses", varieties like "Peek a Boo". These have a similar height to a low-growing floribunda (cluster flowered) but are less woody and have miniature-like cluster flowers. I see a great future for this type of rose both in the garden and as a house plant, and at the moment we have a large range of varieties of our own raising under trials.'

In the United States the large commercial firms, like Jackson and Perkins and Armstrong Nurseries, are producing new miniature roses, but by far the greatest number of new varieties are grown by miniature rose specialists. Some of these are small operations like those of Dee Bennett of California or Ernest Williams of Texas, where only a few thousand seedlings are grown. But small does not mean a lack of quality and from these growers have come varieties like 'Plum Duffy' and 'Hot Shot', 'Dreamglo' and 'Hula Girl'.

Others, like Ralph Moore, work on a large scale and grow many thousands of seedlings each year.

What is probably the largest miniature rose breeding programme in the world operates just north of Boston, where Harmon Saville has Nor' East Miniature Roses. Here about 100 000 seeds are planted each year to produce as many as 80 000 seedlings, all miniatures. Of these perhaps 200 will be selected to grow for another year. The seed is planted out in trays and as it comes through, each seedling is pricked out into its own individual 5 cm pot. Each day, while the seedlings are in flower, Harm goes through the seedling house discarding rejects by the armful. He is now rejecting plants which are better than many which were new twenty years ago. To be released for sale, a new rose must be exceptional in some way, with no weak points. And what is a good rose? 'The bush must be vigorous, but not too tall. The foliage must be healthy. The flowers should have crisp petals and a clear colour. It must repeat quickly. The bloom must not be too large and have a high pointed centre.' More and more of Harm's roses are approaching this standard. Of his newer roses, Harm considers 'Minnie Pearl', 'Acey Deucy' and 'Rainbow's End' among the best. And the top of his list? — 'Spice Drop'.

Over the past forty years miniature roses have come a long way. In the early 1940s, there were only a handful of miniature roses, but in 1954 five new varieties (three from Ralph Moore) came onto the market at the same time. And then in 1973 there were eleven new varieties (ten from Ralph Moore), while in 1983 74 new miniature roses were registered. The number of new varieties produced each year has increased tremendously. Of all new roses registered in 1983, nearly 30 per cent were miniatures. While they represented varieties from seven different countries, 59 came from the United States. Miniature roses have come a long way from a small rose on a window ledge in Switzerland, and there seems to be no reason why they should not go on improving. In coming years we can expect a still greater variety of colours, different forms of growth and greater disease resistance.

PROPAGATION

While wandering through a nursery or garden centre, we sometimes wonder how much is known and appreciated about the time and care that have gone into the production of the thousands of plants that are being offered for sale. From comments we hear, some people seem to think that plants are made on a production line in the same way that shirts or shoes are manufactured, and that if we are out of stock today, a phone call to the warehouse will see more arriving in a day or two. This is not the case.

Some plants are grown from seed, but this does not include roses bought in a garden centre. Only plants which come true to type from seed can be propagated in this way and we have already seen how each miniature rose grown from seed is unique and different in appearance from its parents.

Most miniature rose plants are propagated either from cuttings or by budding. Both methods have their advocates, and both have advantages and disadvantages. From the buyer's point of view, budding produces a large plant more rapidly, but the plant continues to grow more vigorously and may become coarse, with the stem, leaves and flowers losing some of their petiteness. There can also be a risk of suckers. While cuttings take a little longer to grow into larger plants, the final result is more truly miniature and, by suitable trimming, the plant can be encouraged to remain bushy and compact. Budded miniature roses are often sold as bare root plants during the winter while they are dormant. Plants grown from cuttings are usually sold in containers and, while they can be planted into the garden at any time, it is better to do this while they are making rapid growth in the spring or early autumn. (See 'The Miniature Rose Year'.)

Under nursery conditions, miniature rose cuttings are made from flowering stems any time after the bud has begun to swell. These half-ripe cuttings are grown under mist where they are kept constantly moist. They may be bulked up close together or, where space is no problem, planted directly into individual containers. Within a few weeks, roots form and, where necessary,

the cuttings are transplanted into separate containers. The amount of misting is reduced and the small plants, because that is what they now are, are moved out of the propagating area to grow on.

The home gardener can also grow miniature roses from cuttings. Some varieties will grow quite easily while others can be very disappointing. You will need to experiment to find which do best for you. Select a healthy stem where the flower has just faded; it will need to have at least three or four leaves. Make the bottom cut just below a leaf. At the top end, remove the dead flower by cutting just above a leaf. Remove the bottom leaf or two from the stem but make sure that your final cutting has at least two leaves remaining. This cutting is now ready to plant in a container filled with any free-draining mix.

Coarse sand is sometimes used, as are perlite or vermiculite, either alone or in combination with peat. Potting mixes can be satisfactory provided they do not become waterlogged. You will need something that will remain moist without becoming really wet. This is most important if the containers are to be watered or misted frequently. Commercial misting systems are usually controlled by timers or electronic sensors which are designed to keep the cuttings moist but not to overwater them. This ideal is difficult to maintain with a hose or watering can. The cutting should be planted to a depth up to half its length. A little rooting hormone (easily purchased) on the base of the cutting will help it to root. After planting, water the cutting thoroughly. Several cuttings can be planted in the same container, but remember to label them in some way.

Your newly planted cuttings need light to encourage them to grow, but too much sun can dry them out. A bright, humid greenhouse is ideal. You could use a place in the garden where they will be in partial shade but will be seen frequently and can be easily sprinkled as required. You might like to create your own mini-greenhouse. With the cuttings in the container, place a large plastic bag over the cuttings and fasten it with a rubber band or string around the top of the container. A few pieces of stick or loops of wire can be used to support the bag away from the cuttings. The inside of the bag will soon fog up as droplets of moisture form on its surface, but don't worry. This shows that the humidity inside the bag is high and the cuttings are moist. Don't remove the bag when watering is necessary. Place the whole container in a bucket with several centimetres of water and let the moisture soak up from below. When the cuttings show signs of growth, cut the corners off the bag and gradually open up the new plants to the air over several days.

Don't be in too much hurry to transplant your rooted cuttings. They can appear to be growing strongly but still have no root, as the new leaves can be produced from the sap within the cutting. Even when the old leaves fall, as they probably will, don't despair. If the stems still look healthy they may still grow. Depending upon the time of the year when you took your cuttings, it could take up to two or three months before your cuttings are ready to pot up into separate containers.

When is the best time to try growing cuttings? Whenever you can produce the right conditions. For many, that means late summer or early autumn while the weather is still warm and bright but the days are not too hot to dry the cuttings out too quickly. Cuttings made in the autumn can be left in their original containers all winter and will come to no harm, provided they are not allowed to become too dry or too wet.

It is not easy for the home gardener to propagate miniature roses by budding. They are budded on to understock in the same way that bush roses are grown. To do this the grower prepares understock in the late autumn; a form of *R. multiflora* is most often used. Cuttings are made, with the buds, which will be below the ground, removed to prevent the growth of suckers. *R. multiflora* cuttings root easily and grow rapidly and by the next summer will have long shoots a metre and more in length. While the understock is in full growth and the bark lifts easily, a T-cut is made close to the ground level. The miniature rose bud is placed into this and tied

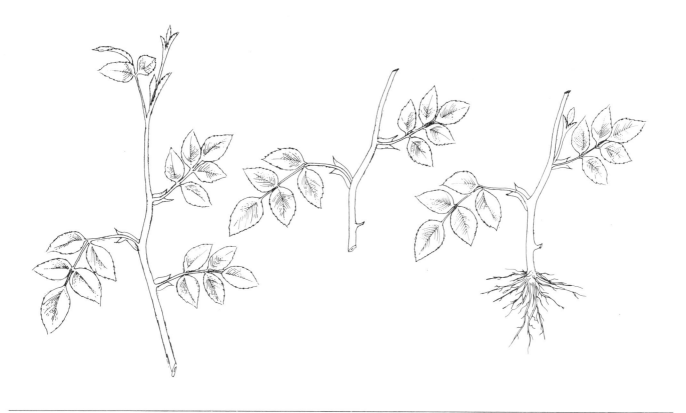

in. During the winter the top *Multiflora* growth is removed close to the tie. In the spring the miniature rose bud begins to grow, and by the end of the summer quite a large plant may have formed. These are lifted during the early winter to be sold as bare root plants or planted into containers for later sale.

Standard miniature roses are produced in the same way, but with them the buds are placed into cuts made at the required height above the ground. Miniature rose standards may only be 45 cm high, but the height can vary. The traditional weeping standard is budded up to 2 m above the ground. Standards are difficult to produce as the long stems dry out easily before they root. As well, each stem must be tied individually to some form of support. They can be most attractive in flower, but it is no wonder that they can be costly.

These are the traditional ways of propagating many plants, including miniature roses. In recent years, experiments have been made with a new method — tissue culture or micropropagation. With tissue culture, a small piece of plant material taken from the tip of a growing shoot is sterilised and placed in an artificial culture medium, usually an agar containing all the necessary nutrients and plant hormones. Under these conditions, the plant tissue multiplies and forms minute 'plantlets' which, at the right stage, can be divided and the process repeated. In this way one piece becomes ten, ten becomes a hundred and a hundred a thousand, in a relatively short space of time. When rooting hormone is added to the growing medium, the plantlets stop dividing and grow roots. All this is done under completely sterile conditions with the plantlets growing in closed containers. When the rooted plantlets are large enough, they are taken from the containers, planted into a sterile mix and hardened off.

Miniature roses can be grown by tissue culture but the process requires special equipment and is not easy. It does, however, mean that many plants can be produced from a small amount of material in a relatively short time. With special varieties in great demand, it may be worthwhile. Perhaps with tissue culture we are getting a glimpse into the future.

USING MINIATURE ROSES

Back in the early 1800s many of the first miniature roses were sold as flowering pot plants. They were small, apparently a little delicate and in need of special attention. (In fact, although miniature roses can be grown successfully in pots, they are not really house plants.) Later the realisation that they are really very hardy meant that they were used in other ways; today most miniature roses in New Zealand are bought as garden plants.

The use of miniature roses as rock garden plants was a logical development with those small early varieties. 'Midget', 'Red Imp', 'Cinderella' and other small-growing plants of the 1940s and early 1950s were on the same scale as many other dwarf rock garden plants and added colour during the summer months after the other spring flowers had finished. Providing colour in the rock garden is still an obvious application of miniature roses but, with the greater range available, they can also be used to add height and interest.

Care must be taken when planting in a rock garden that sufficient depth of soil is left for the roots. Miniature roses will not grow successfully in shallow pockets of soil among the rocks, especially when in full sun. The rocks absorb heat which is reflected back onto the plants, which in turn dehydrate.

Instead of making that raised patch of ground in your garden into a rock garden, consider using a bark mulch to cover the bare ground between your plants. It will help keep the roots cool and moist and, as it weathers, it makes a natural-looking background for your miniature roses. When you grow them in this way, you need space between the plants to show them off. Don't be misled by the size of the plants when they first arrive to be planted. Within a year they could be 30 cm across and they should be planted 45 to 50 cm apart.

Miniature roses can be used successfully as a colourful border to the flower garden. Planted 25 to 30 cm apart, they soon form a continuous line along the edge of your path or lawn. Keep them at least 20 cm back from the lawn or there will be problems later as the plants encroach on the grass. Select your varieties carefully if you want uniformity in height. Don't plant 'Stacey Sue' next to 'Dreamglo', as the former is short while the latter is too tall for any border. Shorter varieties are often better than tall-growing ones. 'Cinderella', 'Dwarfking', 'Marilyn', 'Midget' and 'Little Linda' have similar growth and can all be kept below 30 cm. They could be planted together to give a range of colours. 'Cuddles', 'Red Flush', 'Watercolour' and a host of others all grow to 40 to 50 cm.

Miniature roses make a good border to a rose garden. By planting closer to the edge of the garden and between the other roses, you can add a row of miniature roses and the garden will need very little widening. Again, you must choose your plants carefully. Don't plant tall-growing miniatures in front of short bush roses; you could finish up with miniatures that are taller than your other roses. There may also be colour clashes that you wish to avoid, though some growers feel that contrasting colours give greater emphasis to each variety.

The greater use of evergreens and shrubs in the garden creates another role for miniature roses. Many shrubs are once flowering, and while the foliage may remain attractive during the summer, they really need some colour around them. Plant shrubs as a backdrop to an informal scattering of miniature roses set out in the same way as bedding plants. For the greatest impact, plant them in blocks of one colour. Mass plantings of miniature roses to give banks of colour are sure to become more common. In this type of planting, the shape of the individual flower is less important than the number of flowers and the massed effect. The flowers and plants are often larger and the colours brighter. Imagine the eye-catching splash of colour from a massed display of 'Wanaka', 'Ragtime' or 'Chattem Centennial'.

If you have always wanted your own rose garden but have never had the space, why not create one with miniature roses? There are sufficient varieties and types available to create any effect you wish. It can be formal or informal; it can consist of one plant of a variety or groups of three or four of the same. If you have a leaning towards older roses, choose some of the single and older types. Look for the charm of 'Simplex' or 'Angel Darling' and don't forget the moss roses; 'Dresden Doll' deserves a place in any garden.

Any rose garden needs variation in height to add interest. As well as the differences which can be achieved by changing varieties, special features can be added by the use of standards and miniature climbers. Standards in a rose bed give extra height and at the same time allow the planting of suitable varieties between and beneath them. When people think of climbers, they usually picture them growing against a fence or trellis. Miniature climbers do well when grown this way and are easily managed. Although they will grow 2 m high and at least 1 m across, they need less training than the larger varieties.

For something different, try a miniature such as 'Little Girl' or 'Jeanne Lajoie' as a pillar. Use three 2 m long stakes tied at the top to form a tripod. Make sure the ends are firmly embedded in the ground and about 45 cm apart. Plant your miniature climber between the legs and, as it grows, tie the long shoots to your tripod. In this way you can have a pillar of flowers from the ground up.

Gardeners sometimes need to cover a space such as bare ground between tall-growing shrubs or a bank where little else will grow. Or it may just be an area where they decide to try something different. What is needed is a plant which does not grow upright but hugs the ground and which will, in time, form a mat completely covering the surface. Ivy and other similar trailing plants have frequently been used in this way; but you can do the same thing with miniature roses.

One of the first of the true groundcovers was 'Snow Carpet'. Its small white flowers are not very exciting and in many areas it flowers only in the spring. Also, it doesn't always grow very well until it has had a chance to become well established. But these things didn't matter too much at the time it was introduced. It was new and no other rose was quite like it. 'Snow Carpet' puts out thorny stems which creep along the ground. In our conditions it does not grow very

quickly but it will eventually cover a satisfactory area. Following 'Snow Carpet' has come 'Angelita'. It has a better flower with more petals and often a touch of cream or pink. It also has more reliable growth and forms a better plant more quickly. Of course Sam McGredy is busy trying to achieve further improvements and variations in this type of miniature rose.

From Japan comes 'Nozomi', another ground-cover with long, vigorous trailing shoots which can stretch 2 m or more in a season. The following spring there will be clusters of small pink single flowers along its whole length. It does not form as complete a cover as 'Snow Carpet', but it will quickly cover a large area.

Other roses sometimes used for similar purposes are the 'Cascades' — 'Red Cascade', 'Orange Cascade' and 'Pink Cascade'. They don't crawl, but arch outwards to form a mound a metre or two across. These groundcover miniatures are ideal on a bank or at the top of a wall where they can be seen easily as they drape themselves over the edge.

As well as making excellent garden plants, miniature roses are ideal for containers. For many, the idea of growing a rose in a pot is novel and they wonder how well it can be done. Miniature roses have been grown this way for many years. Roulet discovered his miniature rose growing in a pot and it is almost certain that a hundred years before this the first miniature roses were being treated as house plants.

If you already grow plants successfully in containers you should have no problems with miniature roses, as they are very accommodating plants. Provided you can give them their two essentials, light and moisture, they will grow and bloom profusely. If, however, you are about to begin container growing for the first time, there are some important points to consider.

With the majority of container-grown plants moisture is the most critical factor — too little, and the plant withers, too much, and the root rots and the plant dies. Because of the small volume of material many containers hold, maintaining the correct moisture balance can be a problem. Take a few simple precautions. Consider carefully the type of mix you use in the container. For miniature roses it must be able to retain moisture but also be free draining so that any excess water can drain away quickly. Any good potting mix will do. Such mixes often consist of a mixture of peat, coarse sand and sawdust, although today pulverised bark may partially or even completely replace the peat and sawdust. Make up your own mix if you wish, but do remember to use river sand and, if you use sawdust, make sure that it comes from untreated timber. What about garden soil? We don't recommend it, especially in small containers, but by all means add some to the mix for a large tub. On its own, soil can pack down and become very hard, and if it gets at all dry it can be very difficult to water thoroughly. It is also heavy and a large container filled with soil can be almost impossible to move. Soil may also contain undesirable weed seeds.

Mixes containing large amounts of peat can be difficult to water if they become too dry. The mix shrinks as it dries and pulls away from the edges of the pot, leaving a space. When the plant is watered the water runs quickly out the bottom giving the impression that the mix is completely wet although the centre is still quite dry. The only easy way to saturate the mix when this happens is to stand the whole pot in a container with water coming part way up the sides of the pot. Leave it to soak until the surface of the mix is wet with the moisture drawn up from the bottom.

With all the water that you are going to pour on the top of the pot, there must be some way for the excess to escape. Commercial pots, whether clay or plastic, usually have adequate drainage for miniature roses provided they have been filled with the right mix. You may wish to use a more attractive container than an ordinary pot and many of these have no drainage holes. This problem can be overcome by using a container within a container. Place a few centimetres of pebbles in the bottom of your outer container. Now find another container with drainage holes which fits completely inside the first; a plastic pot is ideal. With your plant in this, place it on the pebbles in the larger container. You can hide the plastic pot by covering the top with potting mix or moss. You must still be careful not to overwater, as the inside pot must not stand in water. The water level should not come above the top of the pebbles.

Things will be easier if you can use large containers as they do not dry out as quickly. Why not plant several miniature roses in the same container? A half barrel could be used to grow three or four plants around the edge with a taller grower or miniature standard in the centre. Ideal along the edge of a terrace, a long tub or trough can hold a number of plants in a row. One of the most attractive and practical ways of growing miniature roses we have seen used wooden troughs but the roses in them were planted in pots. The plots were placed in the trough, the spaces between them filled with old sawdust so that the pots were just hidden. It was an easy matter to rearrange the 'garden' if plants grew unevenly or colours clashed. Old plants could be replaced with new varieties without disturbing those that were to remain. As well as making the whole thing look tidy, the sawdust helped to keep the pots cool and moist. Some of the troughs sat on the edge of a terrace, but others surrounding a small lawn were on legs which raised the plants about 50 cm, so that they could be admired more easily.

In this modern age you can buy a variety of different systems to help overcome your watering problems. As well as a range of sprinklers, there are microtubes and emitters which, when placed one to a pot, drip the water in controlled amounts exactly where it is wanted, although even these can overwater a plant if they are left running too long. You can, however, obtain timers which turn the water off after a set period. With the right mix and the right container your watering problems can be overcome, but please don't put your miniature rose in its container in a place where it will be forgotten and neglected.

Miniature roses in containers can be placed along the edge of a path or used to soften the line of the foundations of the house. Grow miniature roses in a window box. In fact, grow them anywhere as long as they will receive sufficient light and sun, for, without these, they will not flower. This means that they should probably not be grown in the house.

Miniature roses are not really house plants. They can be grown indoors but only if their requirements of light, moisture and humidity can be met. If you have a sunny window, your plants will probably get enough light and watering can maintain the necessary moisture level in the pot, but humidity can still be a problem. Modern homes, especially those with central or electrical heating, have a dry atmosphere and under these conditions, even though the mix around the roots may be damp, the leaves can lose moisture faster than it can be brought up from the roots and the plant withers and dies. If you really want to grow your plants indoors you must provide the necessary humidity. Try placing the pots on a dish or tray containing a centimetre or two of pebbles. The tray catches the excess water as it drains through the pots and the moisture evaporating from the damp pebbles keeps the air around the plants more humid, helping to prevent dehydration.

The easiest way to have flowering miniature roses inside is to grow them outside or in your greenhouse until they begin to flower. At this stage, bring them in and put them in a spot where they will continue to get ample light. When the flowers have finished, cut the plant back, put it outside again and repeat the cycle.

On those dull winter days when you want something to brighten your life, you can have miniature roses in bloom by creating a completely artificial environment with the necessary conditions of warmth, light, moisture and humidity. Some American growers from the northern states with their bitter winters have developed this technique to a high level and light gardens are popular not only for miniature roses. It is easy to provide the necessary warmth for growing indoors. Day temperatures should be from 20°C (68°F), while night temperatures may go down to

10°C (50°F). Moisture and humidity present no problem with trays of damp pebbles and all the necessary light can be provided artificially. Fluorescent light, particularly from special 'grow light' tubes, can be used. You will need a minimum of two tubes, four if there is little natural light. The lights are left on for 16 hours a day, and even then they are kept very close to the plants. Because the tubes do not get very warm they will not do any harm even if the growing plants touch them. Reflectors hung on either side of the plants can increase the efficiency of the lights. Don't try to grow your plants continuously under lights or indoors. Even miniature roses do better with a rest. When they have finished flowering put them outside; you may even want to plant them in the garden. Don't bring them back inside for the next winter before they have had a chance to be dormant for a few weeks as the colder weather arrives.

We have successfully flowered miniature roses in the middle of winter in a cool greenhouse under a double fluorescent light unit. We arranged it so that it was hanging close to the plants but could be raised when they needed care or watering. One of the greatest problems was moisture on the flowers, as this caused petal rot on some varieties with softer petals. Some of the varieties with deeper colours lost brilliance and became washed out. 'Over the Rainbow', for example, became an uninteresting pink. Small, new plants seemed to do best. In fact, the Americans recommend new plants grown from cuttings during the late summer. For us, some of the better varieties grown under lights were 'Cuddles', 'Peachy White', 'White Angel' and 'Magic Carrousel'. 'Mary Marshall' was outstanding; it grew well, remained an attractive colour and the flower kept its shape.

All miniature roses can be grown in containers. Because their root system is more restricted, they do not grow as vigorously or to the same size but, just as in the garden, some will grow larger than others. In smaller containers the short, bushy varieties which produce a lot of flowers are the best. If you want to begin with something that grows well, try 'Wee Man', 'My Valentine',

'Cinderella', 'Little Linda' and 'White Angel'. The list could go on. Once you have begun, don't be afraid to experiment.

Other types are spectacular in a container in the right place. Imagine a small tub on the corner of a terrace. A low, bushy plant spreads across the tub, arching over its edge to form a mound a metre across and covered with flowers. This could be 'Nozomi' or perhaps 'Sugar Elf'. Other varieties like 'Snow Carpet' will creep across the surface and spill over the edge. Developed as a groundcover plant, 'Snow Carpet' makes a good container specimen. We think 'Angelita' is even better. It doesn't creep as much but forms a larger plant more quickly, has a slightly larger flower with more colour and it repeat flowers. Other varieties sometimes spread and sprawl because of the weight of their flowers. Typical of these are 'Orange Honey', 'Green Ice' and 'The Fairy'.

It is a small step from spreading plants in a tub to spreading plants in a hanging basket. Instead of hanging fuchsias or geraniums why not grow miniature roses? They are hardy and do not need extra protection during the cold of the winter. When they become dormant take them down and put them out in a sheltered spot in the garden. They will withstand frost and snow and, with the warmer weather of spring, come once more into growth. Like all plants in baskets, miniature roses in full growth must not be allowed to dry out. If you have hot, dry winds, the roses may need watering twice a day.

These spreading, sprawling plants need very little pruning. Always trim back untidy growth and remove anything that is diseased or dead, but look after those long, spreading branches. Don't worry if they don't produce bloom in their first year. In the spring, a flowering shoot will grow along each branch wherever a leaf meets the stem and the whole plant will be covered in bloom.

Miniature roses in pots require the same basic attention as those in your garden, but a few extras are necessary for the best results. The large quantities of water being used quickly leach soluble nutrients from the mix and some extra feeding is needed. A pinch of dry fertiliser can be scratched into the surface two or three times during the summer; any general fertiliser will do. We find a liquid fertiliser watered on once a week gives excellent results. Always follow the directions carefully and use liquid fertilisers weaker rather than stronger. Don't repeat what happened in one case we were asked about. The plant had died. Had it been fed? Yes, but it had been forgotten for a few weeks so it was given a dose of concentrated liquid fertiliser at full strength to make up for this. No wonder the plant died!

Don't be afraid to trim your plant quite hard after it has flowered. Cutting the flowering shoots back when the flowers are finished will encourage new growth from lower down on the plant and keep it more compact.

We are often asked how long a plant can be left in a container. All we can do is point to some of the plants we are growing. Currently we grow about a thousand stock plants in containers of some sort. Many of these have been growing that way for more than four years, and some have been in containers much longer. The plant of 'Si' which we originally imported back in 1977 has spent all its life in a pot. It began in a 7 cm pot and each year as its roots filled the pot, it was moved up a size until it reached the 15 cm pot in which it has grown for the last four years. This process of starting small and moving up as needed gives a desirable balance between the size of the pot and the plant. 'Si' is a small plant and does well in a small pot. Given ideal conditions, most miniature roses will fill a 10 cm pot in a few months, but with care can be replanted into something larger while in full growth.

Most of our container-grown plants are in 4 litre containers and even the larger, more vigorous varieties seem to do well. During the summer, the plants are watered at least every second day with a soluble fertiliser at a weak strength in the water. After they have flowered, the plants are cut back and at this time a little general purpose fertiliser is scratched into the surface of the mix. This happens several times during the spring and summer. The plants grow out in the open without any protection from sun, rain and frost. With the colder weather of winter, the plants become dormant and during this time they are given a winter prune and tidied up.

At the same time we examine the condition of each plant. Those which seem a little tired are knocked out of their containers and the bottom 2 or 3 cm of mix is removed. A similar quantity of fresh mix is placed in the bottom of the container and the plant is replaced and firmed in. Where plants have sunk down in the container they are treated in the same way with enough new mix to bring them back to the proper depth. This also gives an opportunity to add a slow-acting fertiliser such as blood and bone to the bottom of the pot. Don't raise the plants too high as the water should not run off the top of the pot. All plants have a centimetre or two of old mix scraped off the top and replaced by new mix which has been fortified with a little general fertiliser. That sets the plant up well for its burst of growth in the spring.

When you have several plants in one container it is worth remembering that miniature roses can be moved quite successfully as dormant plants in the winter. You can lift the plants if it becomes necessary to change all or part of the mix and then replant them where required. In this way there seems to be no reason why miniature roses could not be grown in containers indefinitely.

Try miniature roses in containers. They are easy to grow this way and have a multitude of uses, and they can add an exciting new dimension to your gardening.

Miniature roses also make excellent cut flowers for the home. If you wish to use them in this way, select varieties where the blooms grow on longer stems, with few flowers to a stem. There are many among the newer varieties which grow this way.

Treated correctly, they last well. Flowers wilt when they lose moisture more quickly than it can be replaced. To keep them fresh we must slow down this process and keep as much moisture as possible within the bloom. The first essential is to make sure that the plants have been well watered 24 hours before the flowers are picked.

If the blooms are not full of moisture when they are picked there is little chance of them lasting. When you go to the garden to pick your roses take a container holding a few centimetres of warm water and put the cut stems into the water immediately. Experience shows that cut roses take up warm water more quickly than cold. The water should be pleasantly warm to the touch but not too hot or you can scald the stems and the leaves. Later, as the water cools, fill the container with cold water as deep as possible without the blooms getting wet and leave it for several hours in a cool dark place. Miniature roses, in fact any roses, conditioned in this way will last much longer.

It is impossible to write about roses without writing something about rose shows. Anyone who does something well wants to share their skill with others. The singer wants to sing, the farmer wants to parade his stock and the gardener to show his vegetables and flowers. Although there is competition in a rose show and pleasure in winning, the real reason for a show is to share your best flowers with others. If you are growing your miniature roses well, it will not be long before you want to show off your garden, and to compare your best blooms with others.

In the United States, miniature roses have become so popular that even in the biggest rose shows there are often as many miniature roses as there are large roses. In some areas it has become necessary to hold special shows devoted solely to miniature roses. Although we have not yet reached this point in New Zealand, miniature roses are appearing in ever increasing numbers in shows here.

Requirements for shows differ from country to country and, indeed, sometimes from district to district, but no matter where the show is being held, the judges are looking for the same basic qualities. When judging at rose shows in the United States, we found that while the details were not the same as those with which we were familiar, we were selecting the same best roses as the American judges.

What is it that judges look for? The blooms must have good form. Whatever the type of flower or however far open it may be, the petals should lie within a circular outline and the bloom should not be lopsided or misshapen. The colour should be clear and bright, and the flower should sparkle. The flowers must be fresh and the petals must have good substance. This means that they should have a crispness and a lasting quality which is difficult to describe but is easily recognised when seen. It goes without saying that the flowers and foliage should not be damaged by disease or insects.

Perhaps this sounds complicated. It needn't be. The miniature rose you pick to put in a small vase in some special position in your home is probably the same bloom that the judge is looking for as the best in its class. Do remember that the bloom you pick for your vase has a stem and leaves. These are also important for the show. There are no rules about how long the stem should be or how many leaves are shown; rather the judges look to see whether the bloom, stem and foliage of the exhibit are in balance and make a pleasing whole.

With the multitude of community groups, garden clubs, horticultural and rose societies, there is plenty of opportunity to exhibit your miniature roses. Even if they have to be taken some distance, you can pack your choice blooms into a box and move them all around the country. We have successfully taken miniature roses to shows in many parts of New Zealand. This is our procedure.

When we are preparing for a large show where the blooms have to last for several days, we like to cut our roses the evening before the show. Sometimes it is necessary to cut them even earlier. Recently we took miniature roses from Christchurch to a National Rose Show in Hastings, a journey of some hours. The blooms were picked on Thursday evening and put immediately into warm water. When the water had cooled, the containers were filled to the base of the flowers with cold water and put into an old refrigerator we keep especially for our roses. On Friday morning, the blooms were wrapped, each variety separately, with damp newspaper around the base of the stems and then the whole bunch was wrapped in a sheet of paper. They were placed carefully in a box which was closed and sealed when it was filled. They were flown to Hastings that morning. On arrival at the motel, they were unpacked and again placed in containers filled with warm water. Unfortunately there was no refrigerator to keep them in, so we placed the containers in the bath which was partly filled with cold water in order to keep the roses cool. The next morning, they competed successfully with many others at the show. By Sunday, four days after they had been picked, the roses showed no signs of drooping and still appeared fresh.

If you are just going down the road to the local hall, there is no need to go to all this trouble. Pick your roses an hour or two before you need them and give them the warm water treatment. They will carry quite well to your show without special care. If you are going long distances by car, it is, of course, better if you can carry the roses with their stems in water.

You may have never even thought of exhibiting miniature roses from your garden. If the idea appeals to you, give it a try. If you contact your local Rose Society they will be pleased to help you get started.

Even if the idea of exhibiting miniature roses does not appeal, there are still many uses for cut blooms. A small bowl filled with miniature rose buds and perhaps a little greenery takes very little flower arranging skill and brings much pleasure to the elderly or the sick at home or in hospital when you go visiting.

PLANTING AND PRUNING

Planting — where, when and how

Of all the questions we are asked about growing miniature roses, one of the most frequent is, 'Where can I grow them?' The answer is, 'Almost anywhere.' If you try to grow them in a cold, damp place which gets no sun there could be problems, but elsewhere miniature roses will grow and flower. Of course they grow better under some conditions than others.

Like most flowering plants, miniature roses need sun and, if they are to grow and flower well, they should receive at least two or three hours a day. They will grow in full shade, but under these conditions produce more leaf and fewer flowers. Some varieties like a little shade as there is less fading and the colours remain deeper but, even with these varieties, dappled shade from trees or shrubs is better than solid shade from a house or fence. Miniature roses grow well in full sun and tend to flower better when grown in the open.

Any good garden soil will do, but some preparation will help your plants to grow well. When you have selected the place for your plants, cultivate the whole area to at least a good spade's depth. You must remember that while the root system of your new plant is small, within a year or two the roots may have reached a depth of 40 cm and spread a similar distance, provided the ground has been prepared to encourage this root growth. If you have some old compost, you might work some into the soil at this time. Do not use animal manures unless the area is to be left for several weeks before being planted. There is a danger that fresh manures of any kind will burn the tender roots of new plants. Make sure the ground is free from perennial weeds. There is nothing worse than trying to remove a dandelion from the middle of an established miniature rose.

The best time to plant miniature roses depends upon the way in which the plants have been grown. Plants grown from cuttings will probably be bought in bags or containers of some kind. These plants can, in theory, be planted into your garden at any time of the year as their roots will be disturbed very little if they are handled carefully. Experience has shown us that the best results occur if the roses are planted while they are in full growth. Dormant plants put into the ground during the winter are slow to come into growth and are more likely to be lost. Spring is a good time. After the colder temperatures of winter, the warmer weather brings new life surging back into the plants and, because the days are not yet too hot, the soil around the roots does not dry quickly, so that conditions are ideal for the plants to become established in their new situation. Similar conditions exist in the autumn and this too is a good time to establish container-grown plants in your garden. Plants set out during the hottest days of summer will grow successfully, but they must be given extra care to ensure that they do not dry out while their roots are spreading deeper into the cooler, damper soil beneath the surface.

If the plants you buy have been budded, then you have little choice as to when to plant. Occasionally budded miniature roses are supplied as container-grown plants which can be planted in the summer in the same way as those grown from cuttings, but more often budded plants will come to you bare root during the winter months while the plant is dormant. The grower will have taken care that the roots have not dried out.

You do not need to plant your miniature roses immediately provided you take the same care. The easiest way to care for bare root plants before planting is to dig a trench in any vacant piece of garden, unwrap your plants and place their roots in the trench where they can be covered with soil. If the weather is dry, water them in. Budded roses can be safely held in this way for several weeks if the weather is too cold or wet for permanent planting.

Let us suppose that you have prepared your ground and that your plants have arrived in their containers. Don't worry if some seem a little small. Remember they are miniature roses and, provided they are well rooted, they will grow well. In fact, like many other shrubs and trees, small plants transplant better and grow faster than larger plants. And now you are ready to put your plants into the garden. Of course you have thought about just where each plant is to go and perhaps you have marked the position in some way. Before you remove a plant from its container, dig the hole into which it is to go. The hole should be large enough for you to have your hand around the roots of the plant as it goes in. Its depth should allow the surface of the material in the container to be just below the surface of your garden soil. Test the size of the hole while the plant is still in its container and make sure it is right. Loosen the soil in the bottom of the hole. A handful of peat mixed into the soil will help the roots to become established, especially if you have heavy or very sandy soil. Do not put fertiliser into the hole.

Carefully remove the plant from its container. Plastic bags can be cut down the side with a knife and peeled off. Plants in rigid pots can be turned upside down and carefully knocked out into one hand. Make sure the mix in the pot has been kept moist so that it does not fall away from the roots. Place the plant in the prepared hole and fill in the space around it. If the ground is dry, water the plant thoroughly before the hole has been completely filled in. Firm the soil around the plant and level it off. Don't forget to put in a label with your plant. At this stage you know the variety you have just planted, but will you in six months' time? It is nice to be able to name your plants when they are being admired by your friends.

If you have bare root plants, treat them in much the same way, although you will need a larger hole. It should be large enough for the plant to be placed in the hole without cramping the roots. You will probably be able to see on your plant the depth at which it was growing in the nursery. It should be replanted at, or just below, this so that the bud union (the point where the bud was inserted and all the stems branch from) is just above the surface. Spread the roots in the hole and cover with soil. Again, if the soil is dry, water before the hole has been completely filled in. Make sure you firm the soil about the roots so that there are no pockets of air left.

After planting your new miniature roses, watch the weather carefully over the next few weeks.

If there is little or no rain you will need to water to ensure that the ground does not dry out. It need not be soaking wet, but it should be kept moist. It is sometimes forgotten that new plants do not have a deep root system and the top 6 or 7 cm of soil can quickly become quite dry, before we realise it.

Plant standard miniature roses in the same way as other miniature rose plants. They will, however, need a good stake. If the plant is bare root, put the stake into the hole before the plant and make sure the plant is close up against the stake when it goes in. The stake should come well up the stem and, in windy situations, into the head of the plant, so that some of the stronger branches can be tied to the stake for extra support. A little packing between stake and stem will prevent damage by rubbing, as the plant moves in the wind. Stems can be straightened by tying them carefully to the stake in several places, but when they are straight we would suggest that the stem should be tied to the stake only at the top. If you have stormy weather and a tie should break, no harm will be done if the only tie is at the top; the plant will blow over but suffer no real damage. If the top tie breaks and there is a second tie lower down, the stem may break at this point as the plant blows over.

As your new plant begins to grow, check that there are no stems with tips that are not growing. This is especially important with those roses planted as dormant plants in the winter. They are likely to have a number of stems which will have to be cut back to growing buds.

Pruning
More than any other task in the rose garden, pruning is the one that seems to worry growers and dissuades others from starting. It need not be like this, as pruning is really an easy task and gives a great deal of satisfaction when it is completed. There's nothing quite like a well-pruned rose bed. It is difficult to learn how to prune from a book. Having read the instructions carefully and examined any diagrams in close detail, you go out to the garden to begin work

on your own roses and none of them ever bears any resemblance to those that have been described or shown in the book! With this very much in mind, we shall now explain how to prune miniature roses.

Let us begin by saying that no matter how badly or how well you prune, you won't do your miniature roses any harm. And secondly, that it is better to have pruned badly than never to have pruned at all. In order to know better *how* to prune, you need to appreciate *why* you prune.

Most modern roses, and this includes miniatures, produce their blooms on new growth. The stronger this growth, the better the bloom. Left to itself, a rose will grow these new shoots in the spring, flower on their tips and then, its seasonal cycle complete, rest. Some varieties will grow a second crop of new shoots high on the bush, but these are smaller and weaker and the flowers they produce are also much smaller. As the years go by and a branch gets older, the new growth it makes each spring decreases, until eventually growth ceases and the branch dies. When a branch dies or is broken, the rose attempts to replace it with new growth, frequently from low on the bush. If the plant is healthy, this growth is strong and the whole cycle begins again. After a few years, the plant becomes a tangle of old and dying wood with few flowering shoots which produce anything worthwhile. Pruning encourages new growth, thus keeping the plant young and healthy and improving flowering.

Begin pruning by cutting out completely any dead or dying shoots. Always cut back to healthy wood. This will sometimes mean cutting a branch as close as possible to the base of the plant. At the same time, the centre of the bush should be kept open to allow light and air to penetrate and to help prevent the collection of decaying rubbish. Don't let dead flowers and leaves collect in the centre of your bush, as this provides an ideal breeding ground for disease and insects. Remember, when thinning out this growth, to leave the newer stems unless these are so weak that they could not support another flowering stem. New growth can usually be recognised by

its greener stem and smoother bark.

If you've got this far with your pruning, stop for a moment and look at what you now have — a plant with only half the number of shoots growing from the base, but none have been shortened. How hard should these be cut back? There is no easy answer to this question.

How hard a rose is pruned affects the quantity and quality of the flowers. It can also affect the ultimate height of the bush. If you want a mass of flowers in the spring, then just trim back lightly. It won't matter if you leave a lot of short, twiggy stems as they will all produce a flower. Because of the quantity of bloom, most of the stems will be short and thin and the flowers will be unsuitable for picking. Some varieties always present their flowers on short stems and there is little point in pruning these hard. This applies especially to a number of old varieties such as 'Midget' or 'Yellow Bantam', but it is just as true of some of the newer fine-leaved ones like 'Baby Betsy McCall'. With these, pruning consists more of a tidy up with a little thinning and shaping to the required size.

On the other hand, if you want flowers on longer, stronger stems which can be picked, then prune your miniature roses in exactly the same way as bush roses. Remove the twiggy growth. Remember that new growth cannot normally be thicker than the stem from which it comes. Reduce the remaining growth by between a half and a third of its length. Shoots which have had flowers, if not already cut back, must be shortened, as no good shoots will come from the flowering tips.

Always cut just above a bud — not too close or you will damage it, but not too far away or you will leave a stub which will die back. Try to choose a bud pointing in the direction in which you want the plant to grow. As a general rule these should point outwards, as this helps to keep the centre of the bush less crowded, but there are times when the bud at the right height points in the wrong direction. The plant is going to grow just as well even if the shoot grows inwards instead of towards the outside. There will be other times when you want a plant to grow away from a path or from its neighbour. This can be achieved by pruning to an appropriate bud pointing in the required direction.

We have, of course, been talking about pruning established plants. New plants may require very little pruning in their first year. Unless they have made a lot of growth, just remove any twigs that have died, do a little tidying up and leave them to grow.

You cannot keep a rose short by pruning it hard. If it wants to grow tall, it will. On the other hand, if you just trim your bushes lightly, each year a little more is added to their height and they grow beyond their normal limits. Try not pruning a bush at all for a year or two and see what happens.

You can cut a miniature rose to ground level, especially when it is growing on its own roots. In areas of North America with harsh winters, where the temperature consistently remains below zero for days, the bitter cold and winds can kill rose bushes. We have always admired the courage and enthusiasm of rose growers in these conditions, as bush roses must be covered if they are going to live through the winter, and even then a number will die. Under these conditions, miniature roses can be frosted to ground level, a natural pruning, but the plants come back and by mid summer are in full growth again.

When we first began growing miniature roses, we were told to prune them with the hedge shears — 'Just shear them off to the height you want.' I don't think we ever did use hedge shears. Pruned in this way, miniature roses tend to collect leaves and other rubbish among the branches. We prune our miniature roses with the same equipment used for the other roses. Most important is a good pair of secateurs which must be kept sharp. It is possible to buy finer, pointed cutters which make it easier to get between some of the shoots. We don't have to use a saw, but we do wear leather gloves. When you work with a large number of plants it makes life so much more comfortable.

Miniature standards are pruned in the same way as ordinary miniature roses; they are, after all, only miniatures raised off the ground. Because they are budded plants, they will probably give heavier, stronger growth to work with.

Climbing miniature roses need different attention. The same comments apply regarding the removal of dead or diseased material and keeping new shoots in preference to old, but here the similarity ends. The object is to keep the long strong shoots with as much length as possible. These are tied onto your framework or trellis to form the required shape. Frequently they are fanned out in the traditional v-shape and it is easier to do this while the new shoots are growing and they are soft and pliable. When pruning, it is usual to remove only the tips of these shoots.

Flowering shoots which have grown along the length of last year's branches should be shortened back. Leave only one or two buds on the very thinnest. Stronger shoots can be left longer with more buds to grow in the spring. The strongest can be just lightly trimmed and tied in to add to the framework of the plant and fill a gap if this is needed.

People sometimes look at a climber like 'Little Girl' and ask if they can make it into a bush rose by pruning it hard. The answer is, emphatically, 'No'. A healthy climber will always grow long, strong shoots and the more it is cut back, the more it will attempt to compensate with new growth.

Spreading plants like 'Nozomi' or 'Sugar Elf' are really climbers which have not been supported and cannot support themselves. As with other climbers, it is those long shoots which form the framework of the plant and they must be preserved unless you are fortunate enough to have more than you require. If this happens, remove a few of the oldest branches completely. Whether grown in the ground, in a container or as a standard, these plants all require the same type of treatment. Prune them in a similar fashion to climbers. Remove anything that is dead. Leave the long spreading shoots intact and trim back the side growth lightly. Pruning is usually lighter than with climbers and more thin growth is left, as with these varieties we are usually more interested in masses of flowers than in flowers with long stems.

The same is true of groundcover plants. These require very little pruning other than an occasional tidy up to cut out anything dead or messy. Leave anything else so that such plants can form a mat and completely cover the ground.

Don't be afraid to prune. You cannot kill the bush by pruning and if you find that it hasn't worked well one season, you can alter your style the following winter. Don't be worried. Pick up those secateurs and go to it!

PESTS AND DISEASES

We wish this chapter didn't have to be written. We would like to be able to say that miniature roses are not attacked by insect pests and that no diseases affect them. Unfortunately this is not true. The danger is that a chapter on pests and diseases will give the impression that miniature roses have greater problems than other plants. This also is not true. If mildew, greenfly or other pests are troubling your miniature roses, you can be sure they are about in other parts of your garden too.

The easiest way to control pests and diseases is by spraying. If you intend using sprays, get a good sprayer, as it can make the task so much easier. There is a wide range available, so buy one that suits your whole garden, not just your roses; you will want to spray your cabbages and apples, your peaches and lilies, as well as your miniature roses.

The careful gardener will also purchase a few other pieces of equipment. While some spray materials can be quite toxic, they are safe to use if treated with respect. Buy a pair of rubber gloves and a simple face mask. Use them, especially when there is any wind, or when spraying in an enclosed space where fumes may collect. Follow a few other simple precautions. Keep all spray materials well away from children or pets. Read the directions carefully and use only as directed. Never use at a strength greater than that recommended. Don't open and mix concentrated spray material in a shed or room where you can unknowingly breathe in fumes or spray dust. Don't let spray material, especially in its concentrated form, come in contact with the skin. Don't eat or smoke while spraying; you might transfer spray material to your mouth. Do wash yourself and your equipment thoroughly when you finish.

If you haven't done much spraying, here are a few other hints you might find helpful. If you are using powders, mixing them to a paste with a small quantity of water is easier than trying to mix them with large volumes of water. If you have trouble dissolving the powder, try using warm water. Add a spreader-sticker to your spray. This will help give a better cover and lessen the

risk of the spray being washed off by rain. You can buy a commercial product, but your favourite dishwashing liquid used a few drops to the litre will do quite well.

Don't mix more spray than you can use at one time. Never save mixed sprays, especially if two or more materials have been used together. Apart from possible chemical reactions, they will probably become less effective and their use is likely to cause spray damage. Be careful when mixing two different materials together. Sometimes they are incompatible and the mixture could damage your plants. Special care needs to be taken with oils, sulphur and copper sprays.

When you need to spray, choose a still day and preferably one which is not too warm. Early in the morning, before the heat of the day, is ideal. Always try to cover the underside of the leaves, although this is difficult with miniature roses. Some sprayers have a wand with a nozzle which bends upwards and this can be placed underneath the plant. Use as fine a mist as possible. Spray from the base of the plant upwards and let the spray drift down onto the bush. You have a good coverage when the leaves appear damp and the first drop falls to the ground. You certainly don't need to continue until the spray is running off the surface of the leaves.

Two distinct types of spray material are available — insecticides and fungicides. As the names suggest, insecticides control insects, while fungicides control fungus diseases. One will not do the work of the other. New materials are being introduced all the time and the sprays we mention here may be replaced by something better within a short time. Some of the newer spray materials are systemic, that is, they are absorbed into the plant's system where their effect lasts longer. Others are contact sprays and must touch the pest to be effective. Most sprays tend to be preventative and must be used before, not after, an attack takes place. Nothing can restore leaves damaged by mildew or replace pieces chewed out by a caterpillar.

We have tried to keep the list of sprays as simple as possible. You may hear horticulturists speak of others, but they can often be bought only in quantities too large for the home garden, or their use may be restricted to registered growers. These we have not included. We give the material first, then, if applicable, the trade name under which it is sold in New Zealand. We describe only those that have worked for us; this does not mean that there are not others which are just as good. Ask about the products available in your area.

Insecticides

Carbaryl. A contact spray for the control of caterpillars and other chewing insects. Safe to use on roses, vegetables and fruit trees. As it is highly toxic to bees, don't use it on plants in flower if they are likely to be visited by bees.

Maldison (Malathion). A contact spray for the control of greenfly and sucking insects. Apply once, then use again three days later for a complete kill. Safe for general garden use, but follow directions carefully.

Acephate (Orthene). One of the newer systemic insecticides. This is the active insecticide in several combinations such as Shield and Saprene. Will control most insect pests such as caterpillar and greenfly, but not mites. Safe for general use, but do not use on vegetables close to harvest.

Tetradifon (Tedion). A contact spray for mites.

Dicofol (Kelthane). A contact spray for mites. It is included in several general spray mixes.

Fungicides

Captan. An older general purpose protective fungicide.

Lime sulphur. Use as a clean up spray when the plants are dormant. Do not use when the plants are in leaf — it will remove the leaves just as it will remove paint on the house if it comes in contact with it.

Oil sprays. An alternative winter clean up spray. Unless used with great care, oil sprays applied in warm weather can mark foliage badly. Use oil sprays with —

Copper spray. (This is sold under a variety of trade names. Two of the most common forms are copper oxychloride and Bordeaux mixture). Controls a wide range of fungus diseases, but has been replaced for all but winter sprays by —

Mancozeb. One of the older dithane sprays used as a protectorant for many fungus diseases. It is still effective, but has dropped out of favour with many growers as it leaves a white powdery residue on the foliage.

Benomyl (Benlate). One of the first systemic fungicides which gives protection against mildew, blackspot and botrytis.

Triforine (Saprol). Partially systemic, it gives good control of powdery mildew, blackspot and rust. It is included in Shield and Saprene.

Chlorothalonil (Bravo). One of the newer general purpose fungicides.

While a general spray programme will control pests and diseases, it is helpful if you can recognise the more common problems. You can then check that you are using the right material for the control of your specific problem.

Pests

Greenfly (Aphids). Soft-bodied, green or reddish insects which suck the sap. They appear as clusters of tiny insects on buds and young shoots. Spray with Orthene.

Caterpillar. A variety of caterpillars attack roses. The edges of the leaves may be eaten or the young leaves stuck together. Spray with Orthene.

Froghopper (Cuckoo spit). This has only recently appeared in our garden and is present only in the spring. We were not sure at first what it was, and still refer to it as 'spittle bug' because that is what it looks like. Because the hopper is hidden in the froth, it is difficult to spray effectively. A strong spray with the hose is necessary to clear the froth away and this may be all you decide to do — the hopper appears to do little harm although the froth looks terrible!

Beetles. Several types of beetle, including the grass grub beetle, can chew pieces from the edges of your rose leaves and blooms. As they can fly in from neighbouring pastures, parks and lawns, you cannot cure the problem by treating your own lawn. Spray with carbaryl or one of the new synthetic pyrethroid sprays when the beetles are flying in the spring and early summer.

Scale insects. These attack woody plants and while we have never seen them on miniature roses, they may be a problem in some areas. They appear as small, crusty scales on old and neglected wood. Removing old wood when you prune, followed by a spray of winter oil, will control any outbreak of scale. Both Orthene and carbaryl will give some control during the growing season.

Mites. In hot, dry weather mites can be the greatest insect problem of miniature roses and roses in general. You may know them as red spider, as the European mite is red. A greater problem on roses is the two-spotted mite. You won't notice them on your plants, as they are so small that they cannot easily be seen without a magnifying glass. If the leaves, particularly the lower leaves, develop a yellow-bronze mottling, check the underside of the leaves. This is where the mites will be found. Control is difficult as, to be effective, the underside of the leaves must be sprayed. This can be rather awkward with miniature roses as the leaves are often close to the ground. Although special sprays are available for mites, these pests can, unfortunately, develop a resistance to a particular spray if it is used too frequently. Kelthane and Tedion are the two most commonly used by the home gardener. Use one as soon as the first suspicion of mites appear. If it is a bad attack, repeat seven days later.

If mites reappear, change your spray material for the next treatment. Mites do not like water. Although it will not give complete control, frequent use of the hose to wash the underside of the leaves as you water will help.

Apart from insects, small animals such as opossums and rabbits can be pests in some areas. They will, of course, eat most things that grow and new rose shoots seem to be a delicacy. Growers with opossum trouble tell us that, invariably, the best bloom of the season is eaten just as it is about to open. Various deterrents are tried, such as cotton threaded around the plants to annoy the animals. We have also heard of an electric fence being erected around the rose beds 15 cm above the ground. It is possible to buy a chemical deterrent, but we cannot comment on its effectiveness because we are fortunate in not having this problem. If animal pests are a problem for you, you might like to try it.

Fungus diseases

Powdery mildew. When growers talk of mildew, this is what they are usually referring to. It especially affects young shoots, leaves and buds which become distorted and covered with a white powdery substance. Once they have got to this stage, nothing can be done to help them. Use Benlate, Saprol or Bravo as a preventative and control. Prune off badly affected shoots.

Blackspot. This disease is worse in warm, humid areas. Circular black spots which look like little ink drops on blotting paper appear on the leaves, which will later turn yellow and usually drop off. If blackspot is a problem, be sure to use a copper spray during the winter, followed by Benlate, Saprol or Bravo during the growing season.

Rust. Rust looks just like its name. Found mainly on the underside of the leaves, it shows as powdery orange spots which later turn black. With bad attacks, it will show through to the top surface of the leaf as yellow stippling. Once again, use Saprol or Bravo as a control.

Downy mildew. Another disease of warm, humid areas, this is not a great problem with miniature roses. It appears as patches of purple greyish mould on the underside of the leaf with the top surfaces showing a matching yellow. The leaves will usually drop off. Spray with Bravo or mancozeb.

Die back. This is not really a disease but something which occurs for a variety of reasons. It may be the result of damage through faulty pruning cuts, it may be because of the frosting of a soft new growth, or it may just be associated with the natural aging of the plant. It does not normally require special spraying. Remove die back as you see it, remembering always to cut back to healthy wood.

Sometimes a plant just dies for no apparent reason. It may be because of a wandering cat or dog, or there may be no real reason. We see it in the nursery. In a batch of plants all of the same variety and treated identically, one will just wither and die. If this happens in your garden, remove the dead plant, taking some of the soil with it in case the cause is in the soil. Once there is fresh soil back in the hole, it is usually safe to replace your plant.

If you want a simple spray programme to follow, we suggest the following. Use a winter spray of oil and copper. Follow this in the early spring with a combination of Saprol and Orthene. This can be bought as Shield or Saprene. Repeat this monthly. If you are in a disease-prone area or you become interested in entering your roses in shows and wish them to be without blemish, you will want to spray more frequently. Add something for mites as necessary.

If you think this is a lot of spraying or you dislike the use of chemicals in the garden, you can still grow miniature roses. Well-grown plants and the frequent use of the hose will give healthy plants. They may not be disease free, but they will be able to withstand attacks. It will help to remove infected leaves as soon as they are seen. The thumb and finger pressed firmly together are very effective in the control of many insect pests.

We should all practise good garden hygiene and keep the garden free from decaying leaves and flowers which harbour diseases and pests. We can encourage birds into the garden. We frequently see small birds like waxeyes and finches searching through the roses for a tasty titbit. Natural insect predators also help and we should welcome the ladybird and praying mantis. Science is also helping here. Many orchardists now use a mite to control mites; yes, there is a predatory mite which devours the two-spotted mite. Remember, it is no good hoping for healthy miniature roses if the rest of your garden is full of pests and diseases.

THE MINIATURE ROSE YEAR

If those first New Years had been southern festivals and the date determined by New Zealand conditions, we would surely celebrate New Year in July. While this is the coldest time of the year, it is also the time which suggests a new beginning. As the days gradually become a little longer, the spring-flowering bulbs push upwards and the buds of the earliest blossoms begin to swell, reminding us that spring is not far away. It is this time that we regard as the beginning of the miniature rose year.

Winter

Growth will have stopped and in the colder parts of the country roses will have lost their leaves. This time, while they are dormant, is the best time to prune your roses. While pruning, check the ties on your climbers and standards and renew them as necessary. After pruning, tidy up the rose beds. Clean up any rubbish and fallen leaves. Get rid of any weeds. It won't hurt the plant if you need to loosen it to remove a stubborn weed provided the plant is firmed back into the soil afterwards.

Spray your roses and the surrounding ground with a good fungicide. Many growers regard this winter preventative spray as the most important of the year, if diseases are to be controlled.

If you haven't already done so, prepare the area for any new miniature roses you intend planting in the spring. Existing rose beds can be re-organised, if necessary, as even mature plants can be safely moved while dormant. Dig them out carefully with as much root as possible and replant them in their new position in the same way that you would a new, bare root plant.

Remember to check any miniatures in containers. They, too, need pruning and weeding. They can safely be left outside for the winter, but would be better with some protection from the worst of the winter winds and rain. Reduce the watering of plants being grown under cover, but check that they remain damp. Even in the middle of winter they should not be allowed to dry out completely.

Do not feed your miniature roses during the winter.

Spring

Miniature roses in containers should be top-dressed and given a thorough soaking as soon as new growth is seen. They will need to be kept moist for their best growth. If you have a glasshouse or sheltered porch, bring some of your plants under cover and they will start flowering early in the spring.

Early spring is the best time of the year to plant new miniature roses. As soon as the ground warms up sufficiently to encourage root growth, new miniature roses can be put in. If you are uncertain about which varieties to purchase, you can safely delay buying and planting until spring blooms appear. See them in flower and then buy container-grown plants to put into the places you have prepared. Newly planted miniature roses must not be allowed to become too dry.

At the first sign of new leaves give your established miniature roses a little fertiliser.

The soft, new growth of spring is the favourite dish of many garden pests. Keep a watch for greenfly and spray with the correct insecticide as soon as they appear. Watch, too, for small caterpillars feeding on the growing tips of the expanding buds. Use a general fungicide with the insecticide. In this way one spray can serve two purposes.

Growth is very fast at this time of the year and growing plants must be kept moist. Watch for dry spells and use the hose when these occur.

As miniature roses flower, cut off the spent blooms.

Watch for and visit the spring rose show in your area, as this is a good chance to see new and unfamiliar varieties; most shows include miniature roses. Don't be afraid to ask for advice. That magnificent bloom that you see may not be the best for your area.

Summer

Your miniature roses should be in full bloom. Even if neglected, they will continue to flower through the summer but, with a little extra attention, they will do much better. Water is very important at this time. More than anything else, roses require water through the heat of the summer if they are to do well. Miniature roses are no exception. They need 2 or 3 cm of water a week and if it does not come naturally, you will need to use the hose. With established plants, soak the ground thoroughly at least once a week. As the water soaks into the ground, the roots are encouraged to follow it downwards and the plant will still get moisture even when the surface soil dries out. On the other hand, frequent light sprinkling keeps the surface moist while the soil below can become quite dry. The roots never penetrate to any great depth and the plants can suffer in hot dry weather. New plants will require more frequent watering as they have had no chance to develop a deep root system. The ground should be kept damp during this first summer.

Well-grown plants will be covered with flowers. Left to themselves, they will flower again, but not as well and only on short stems. You will get a lot more flower if the flowering shoots are cut back to just above a leaf, after the flowers drop. Instead of a few flowers high on the stem, the plant will branch from lower down and produce stronger stems and more flowers. At the same time, the plant remains more bushy and compact.

During the summer, many of your miniature roses will produce strong new shoots, frequently from close to the base of the plant. On strong, tall-growing varieties, these can reach a height of 70 or 80 cm. At their tips these shoots produce multiple branches with clusters of flowers often high above the rest of the bush. When the flowers have dropped, these shoots should be trimmed back, lightly if you wish to increase the height of the plant, more heavily if you wish to keep the height down. Some growers prefer to pinch the tops out of these strong shoots before they become too tall.

If you are growing climbers, it is these strong shoots which form the framework of your climbing rose. In one season they can grow 2 m long and will need supporting and tying as they grow. Do not cut these back. They may not flower the first year, but don't be disappointed. In following years

they will amply repay the care you give them.

Miniature roses do not need a lot of fertiliser; too much is worse than too little. A little general fertiliser may be sprinkled around the plants whenever they are trimmed back.

The warmth of summer brings out the bugs and encourages diseases. If your miniature roses are growing well and have been trimmed back to encourage new growth, they will survive, but a little spraying will keep them at their best. Always remember that prevention is easier than cure. Watch for the first signs of insect pests or disease and use the appropriate spray material.

Plants in pots and containers need special attention at this time of the year, as they can dry out so easily. They need watering every day and sometimes twice a day if you have hot, drying winds. Remember that with all this watering they will also need some feeding. If the plants do get dry and the leaves shrivel, don't despair. Trim them back, soak the plant thoroughly — put it in a bucket of water — and within a few weeks you could have your plants once again in full leaf.

This is a good time to see miniature roses in flower. Visit local gardens and nurseries to check new varieties. This is the best way to decide upon the varieties you wish to add to your garden.

Container-grown plants can still be planted into the garden, but they will need special attention while they are becoming established. They must not get dry.

Autumn
Growth in your roses will be slowing down, but if you have been trimming back they will still be in flower. This is a good time of the year for your miniature roses. With the cooler weather, they need less watering, the sun is not severe and the colours are deeper as the blooms do not fade.

As the cooler weather approaches, there is no need to continue removing the dead flowers. Even in warm and sheltered areas it is unlikely that new growth at this time of the year will produce flowers. Don't give further fertiliser. Let your plants harden off so that they are better prepared for the winter.

Don't neglect your spray programme. The cooler nights encourage the development of powdery mildew, and greenfly also seem to prefer the cool of autumn to the heat of summer. Cool weather, heavy dews and damp mornings are ideal conditions for botrytis which appears as a grey-brown mould on dying flowers, dead leaves and soft new shoots. The recommended spray programme will control this.

Healthy, disease-free plants, especially those growing in a warm spot, will hold their leaves well into the winter. Some, in fact, will be almost evergreen. If lightly pruned in the autumn, these plants will come into flower early in the spring, as much as a month earlier than roses growing in the open.

Plants growing in containers and trimmed back late in summer or early in the autumn will make new growth if they have been kept in a warm, sheltered place. Provided they receive enough sun, flower buds will be forming. Keep them in a sheltered place and they will flower through into the winter.

If you are planning a new area for miniature roses, autumn is a good time to get it ready. While the weather is still warm, it is also a good time to plant container-grown roses. There is less risk of them drying out and while they are still in growth there is a chance for the roots to become established before the cold of the winter.

TOMORROW

There are not many things you can be certain about these days, but even as we began writing this book we knew there was one thing of which we could be sure. Before we finished, new ideas about miniature roses would emerge and new varieties would be released. There is no real end to what could be written. We already hear of new miniatures from Ralph Moore — 'Earthquake' and 'Why Not' and a new hanging basket variety, 'Sweet Chariot'. The story of miniature roses goes on. We can all see these developments here as more and more miniature roses appear in ever-increasing types and varieties.

Miniature roses are here to stay. They have come too far to once again be lost as fashion and interests change. Each day they grow in popularity as more and more people discover their beauty and their usefulness. As gardens are becoming smaller, miniature roses are being recognised as ideal garden plants, since several can easily be grown in a small space to give a variety of colours and forms. And for those who really become interested, they are collectable. It doesn't require too much room to grow a collection of twenty, or fifty or a hundred different varieties. And miniature roses will continue to improve. It has taken the big roses in our gardens more than a hundred years to reach their present place, whereas our modern miniature roses have been around for less than half this time.

We don't have a crystal ball, but it is interesting to predict the next few steps into the future. We can expect a continuing improvement in the health and vigour of miniature roses. Flower form will be improved and, for those who want them, there will be more with good hybrid tea form. Many of these are likely to grow naturally one bloom to a stem. The florist and flower arranger will realise the full potential of these varieties and they will be grown increasingly as cut flowers. The introduction of compact plants with prominent flowers means we are likely to see more miniature roses as flowering pot plants. We will see more climbing miniatures, more miniatures as flowering plants in containers, more miniatures as groundcovers. While there will always be small miniatures, there is likely to be a greater number of miniatures with larger flowers, the 'in-between' roses. The distinction between floribunda and miniature roses will become increasingly blurred. It doesn't matter whether the rose is called a miniature or a floribunda, a patio rose or a miniflora. We will continue to choose roses for what they are, how they grow and whether they suit our purpose, not for what they are called.

We hear whispers of things to come — a groundcover rose with pink flowers; from Japan a 'blue' miniature rose (a blue-mauve similar in colour to 'Katherine Mansfield' or 'Blue Moon'); and a black miniature rose! Well, not really black, but a red so deep that the bud appears almost black. We must wait to see these, but in the meantime the anticipation only adds to the pleasure and excitement of growing miniature roses.

JEANNE LAJOIE

LIST OF VARIETIES

One of the important responsibilities of any person selling plants is to ensure that when they leave the nursery they are correctly named. Although great care is taken, mistakes do sometimes happen and an individual plant is sent out with the wrong name. Of greater concern is the propagation of plants from wrongly named stock. If a newly-imported variety is wrongly named, or in some way names are muddled in transit, all subsequent plants grown and sold will carry this incorrect name.

Soon after we became interested in miniature roses we bought four new varieties. Among them was a miniature rose labelled 'Tea Party'. As it grew and flowered it seemed at first to be true to description; it was certainly the correct colour. But then it began to grow long shoots like a climber and this wasn't expected, as 'Tea Party' is described as a short, compact plant. Ralph Moore was coming to visit us the next spring and as he had bred 'Tea Party', we carefully tied up the shoots which were by then a metre long. When he arrived, Ralph needed only one look. 'That's not "Tea Party". It's "Little Girl".' And that made sense as 'Little Girl' is a climbing miniature rose. We still grow 'Little Girl' which is a marvellous little rose. We've never grown 'Tea Party', but it continues to be sold, and often it turns out to be a climber!

It is not possible in a book of this type to include every miniature rose. We have tried to show as many as possible of those currently grown in New Zealand and, thanks to the cooperation of growers both here and overseas, several of the new varieties which should become available within the next few years. We have also included a few varieties which are of historical importance and one or two others which are not really miniatures.

We have been as careful as possible with the names and descriptions of the varieties listed and illustrated. Wherever possible we have relied upon our observation of plants growing on our property and describe and picture them as we know them. In some ways it is those we know best that have been the greatest problem. With these we are most aware of the variations of form from bud to open flower. It has been difficult to find one photograph which suggests the changes in colour that can occur with the weather and the seasons. We hope you will remember these variations when you look at the picture of the miniature rose and won't say, 'But mine doesn't look like that!'

The descriptions all follow the same form: the name, followed by any trade name, and then the raiser, the year in which the variety was introduced and, whenever possible, the parents (seed parent x pollen parent). Unless size is specifically mentioned, we are talking about an average-sized flower on an average-sized bush.

AMY'S DELIGHT

ACADEMY (Macgutsy)
McGredy 1982 Anytime x Matangi

A 'hand-painted' miniature, with eye-catching cream blooms edged with orange-pink. The medium height plant flowers early and repeats quickly.

ADA PERRY
Bennett 1978 Little Darling x unnamed Coral Treasure seedling

The coral pink blooms are a good exhibition shape and hold well at the three-quarter open stage. It is a little slower to repeat flower than some varieties, but the flowers are well worth waiting for. The foliage is strong and disease resistant. The plant has an open shape, and is of medium height.

AMY'S DELIGHT
Williams 1980 Little Darling x Little Chief

This small to medium-sized plant has many rosette flowers in a clear medium pink.

ANDREA
Moore 1978 Little Darling x seedling

Shapely buds open to larger flowers of deep pink edged with silver. The plant is vigorous and attracts attention in the garden, where it can grow quite tall if it is well looked after. Mildew can be a problem if it is planted in too sheltered a spot.

ANDREA

ANGEL DARLING

ANGEL DUST

ANGELITA (Macangeli)

ANGEL DARLING
Moore 1976 Little Chief x Angel Face

A most attractive semi-single lavender flower with ruffled edged petals. The yellow stamens are prominent and stay a good colour for a long time. Strong basal shoots produce quite large clusters of flowers, which rise above the medium height plant. These should be cut back when the flowers fade. The lavender colour may be more maroon if the plant is in full sun all day during the summer.

ANGEL DUST
Bennett 1978 Magic Carrousel x Magic Carrousel

Our plants of this variety are very bushy and of medium height, and carry many flowers at one time. These flowers are white with a delicate blush of pink on the edge of the petals.

ANGELITA (Macangeli)
McGredy 1982 Moana x Snow Carpet

The creamy-white-tinged pink flowers of this healthy groundcover plant repeat quickly and make a most attractive display, whether planted to smother weeds, tumbling over walls or down banks, or as a container or hanging basket plant. A very good variety of its type.

ANITA CHARLES (Mornita)   ANN MOORE (Morberg)

ANITA CHARLES (Mornita)
Moore 1981 Golden Glow x Over
The Rainbow

The pointed buds open into well-
shaped flowers with reflexing
petals of coral red and show a gold
reverse which goes almost into a
tan colour. A most unusual colour
combination.

ANN MOORE (Morberg)
Moore 1981 Little Darling x Fire
Princess

As this is the rose Ralph Moore has
chosen to name after his wife, it
must be an outstanding variety.
The flowers have many petals of a
deep orange-red, and are long
lasting. The bush is healthy and
vigorous.

AVANDEL

ANTIQUE ROSE (Morcara)
Moore 1980 Baccara x Little Chief

This variety has rose pink larger
flowers with many petals. It is free
flowering, and the flowers last well
when picked.

ANYTIME
McGredy 1973 New Penny x
Elizabeth of Glamis

The semi-single (twelve petals)
flowers on this vigorous free-
blooming plant are salmon-orange
with an almost metallic look. The
foliage is very dark.

AVANDEL
Moore 1977 Little Darling x New
Penny

A beautiful blend of pink, yellow
and peach colouring with an
attractive bud. The flower opens
flat, and can lose colour as it ages.
The dark leathery foliage is disease
resistant. A medium height plant
with occasional strong basal shoots
which will have large numbers of
flowers on the top. These should be
cut back when the flowers fade.

BABY BETSY McCALL

BABY DARLING

BABY KATIE

BABY BETSY McCALL
Morey 1960 Cécile Brunner x Rosy Jewel

This is a rose we recommend for rock gardens, as it is a short bushy plant with many flowers repeating quickly. The dainty pale pink and cream flowers are borne on stems long enough for cutting. Although the petals appear fragile, they are not affected by the weather.

BABY DARLING
Moore 1964 Little Darling x Magic Wand

This older variety still attracts attention when in flower. It has apricot-pink flowers, sometimes deepening almost to orange, with a well-shaped bud. The stems are sometimes not strong enough to hold the flowers upright, but as the plant matures it usually grows out of this.

BABY KATIE
Saville 1978 Sheri Anne x Watercolour

The exhibition form flowers on this vigorous healthy plant are a delightful blend of pink and cream, and borne one to a stem or in larger heads. It appears to be disease resistant, and repeats quickly. In our opinion, it is one of the more promising new varieties we have in our garden, especially for the keen exhibitor.

BABY MASQUERADE

BABY MASQUERADE
Tantau 1956 Tom Thumb x
Masquerade

No garden is complete without this
free-blooming rose. Its blend of
orange, pink and yellow, with the
aging flowers deepening to a dull
red, is most distinctive and it is so
easy to grow that it is ideal for the
first-time gardener.

BABY PINOCCHIO
Moore 1967 Golden Glow x Little
Buckaroo

The well-shaped flowers are a pink
and yellow blend. As the flower
ages, the outside petals go a deeper
pink while the centre of the flower
stays a deep cream. The bush is
quite large, and has no disease
problems.

BAMBINO
Dot 1953 Pink sport of Perla de
Alcanada

This small plant bears many
flowers in a soft medium pink. It
looks good in a rock garden, or used
as a flowering plant in a small
container.

BABY PINOCCHIO

BAMBINO

BEAUTIFUL DOLL

BIG JOHN

BEAUTY SECRET

BIT O' SPRING

BEAUTIFUL DOLL
Jolly 1982 Unnamed seedling x Zinger

The larger flowers of soft salmon pink look promising for exhibitors, as well as looking good in the garden.

BEAUTY SECRET
Moore 1965 Little Darling x Magic Wand

The bright medium red semi-double flowers have very long pointed buds, and are very fragrant. The plant is vigorous and upright. This variety picks well, and is a favourite for use as a buttonhole rose.

BIG JOHN
Williams 1979 Starburst x Over The Rainbow

Deep red exhibition form blooms with a velvety texture make this a desirable rose, but the flowers are big, as the name suggests. The colour holds well in all weathers.

BIT O' SPRING
Williams 1980 Tom Brown x Golden Angel

Hybrid tea-shaped flowers of peach pink with a blended yellow-pink reverse. There are many blooms on a healthy plant.

BOJANGLES (Jacsun)

BIT O' SUNSHINE

BORN FREE

CALGOLD

CALICO DOLL

BIT O' SUNSHINE
Moore 1958 Copper Glow x Zee

A bush of medium height which bears flowers of bright buttercup yellow with 18 to 20 petals.

BOJANGLES (Jacsun)
Warriner 1983 Spanish Sun x Calgold

Just 40 to 60 cm tall, this reliable clear, lemon-yellow rose grows on a full, well-formed bush. It flowers in clusters all season long.

BO PEEP
De Vink 1950 Cécile Brunner x Tom Thumb

Small medium pink informal flowers borne singly and several together on a bushy short plant.

BORN FREE
Moore 1978 Red Pinocchio x Little Chief

The flowers are a brilliant orange-red, and the plant can be most spectacular when in full bloom. The flowers are at their best when fully open and the stamens stay yellow for a long time. Highly recommended for garden display.

BUTTONHOLE ROSE
See 'Cécile Brunner'.

CALGOLD
Moore 1977 Golden Glow x Peachy White

The double blooms are a deep clear yellow, and complement the dark green glossy foliage. The bush is moderately compact and there are plenty of flowers.

CALICO DOLL
Saville 1979 Rise 'n Shine x Glenfiddich

Semi-double flowers of an orange-yellow blend which are sometimes striped. A healthy medium-sized plant.

CANDY CANE

CANDY CANE
Moore 1958 Seedling x Zee

The semi-double flowers of pink to light red with white stripes are borne in loose clusters on a climbing plant which will grow up to 1.5 m tall.

CANDY PINK
Moore 1969 (*R. wichuraiana* x Floradora) x (Oakington Ruby x Floradora)

A vigorous, tall plant, with small double light pink flowers and fine leathery foliage.

CARELESS MOMENT
Williams 1977 Little Darling x Over The Rainbow

The official description of this small-flowered rose is white edged with pink, but we find that the overall impression is more deep pink than white, especially as the flowers age. The foliage is dark and glossy, and the plant is vigorous and upright, with many flowers.

CARMELA
Moore 1981

Semi-double blooms of an unusual caramel-orange, revealing a bright yellow centre when fully open. The buds are mossed and there are many flowers on a short compact bush.

CANDY PINK

CARELESS MOMENT

CARNIVAL PARADE

CAROL JEAN

CÉCILE BRUNNER

CARNIVAL GLASS
Williams 1979 Seedling x Over The Rainbow

A beautiful blend of pastel orange-yellow with a yellow reverse, the flowers are a good hybrid tea shape.

CARNIVAL PARADE
Williams 1978 Starburst x Over The Rainbow

Very double flowers of golden yellow with red blending. The plant can get some mildew if growing in a sheltered position.

CAROL JEAN
Moore 1977 Pinocchio x Little Chief

A compact plant profusely covered with small deep pink flowers and buds, borne in large clusters on strong stems.

CÉCILE BRUNNER
Pernet-Ducher 1881 Probably a polyantha x Mme de Tartas

Sometimes known as the 'Sweetheart' or 'Buttonhole Rose', this is not a miniature but a polyantha. The pastel pink flowers are often no larger than those of a miniature rose but the leaves, stems, internodes and form of growth quickly indicate that it does not belong. As can be seen, the climbing form can easily stretch 10 m along a fence.

CÉCILE LENS (Lencil)

CÉCILE LENS (Lencil)
Lens 1983

The light pink buds are very shapely, but open quickly into loose semi-double flowers. We cannot find the parentage of this variety, but suspect 'Cécile Brunner' to be somewhere in the background.

CENTER GOLD
Saville 1981 Rise 'n Shine x Kiskadee

The vigorous plant is upright and compact. It carries a mass of deep golden yellow flowers with an occasional white bloom. It is almost continuously in flower.

CHARMGLO
Williams 1980 Unnamed seedling x Over The Rainbow

The well-shaped flowers of deep cream have pink to red edges on the petals. The plant is bushy and compact.

CHATTEM CENTENNIAL
Jolly 1979 Orange Sensation x Zinger

The brilliant blooms of a luminous orange-red on an upright bush make a spectacular focal point in any garden. This variety is good as a specimen plant, in a container on the patio, or in a massed planting in the garden.

CHEERS (Savalot)
Saville 1984 Poker Chip x Zinger

Many well-formed blooms of orange-red with a cream reverse cover the compact bushy plant. In cooler weather the flowers deepen in colour.

CHARMGLO

CHATTEM CENTENNIAL

CHIPPER

CHRISTINE WEINERT

CLARET

CHIPPER
Meilland 1966 (Dany Robin seedling x Fire King) x Perla de Montserrat

This bushy, moderately compact plant has many coral pink flowers of a good shape. We have found this variety looks very good when the weather is hot and dry in the middle of the summer.

CHOO CHOO CENTENNIAL
Jolly 1980 Rise 'n Shine x Grand Opera

Well-shaped flowers of a clear medium pink edged with a paler pink are borne in profusion on a healthy vigorous plant. The foliage is light green and the whole effect is very pleasing.

CHRISTINE WEINERT
Moore 1976 (Little Darling x Eleanor) x (Little Darling x Eleanor)

The semi-double orange-red flowers are borne singly and several together, on a moderately compact plant. According to some references, the plant does better with extra iron.

CINDERELLA
De Vink 1953 Cécile Brunner x Tom Thumb

The very double rosette-shaped flowers are white with a delicate touch of pink, particularly in cool weather. The bushy compact plant has fine foliage and is disease resistant.

CLARET
Saville 1978 Little Chief x Little Chief

This variety has unusual wine-coloured clusters of flowers on a small bush with fine foliage. Each spray of flowers is like a miniature bouquet.

CLOUD NINE (Jaclite)
Warriner 1984 Bon Bon x Calgold

As this variety grows, it spreads up to 40 cm across. The well-formed lemon yellow flowers are semi-double and medium sized.

COLIBRI
Meilland 1958 Goldilocks x Perla de Montserrat

The flowers are a yellow-orange with the deepest colour showing in the cooler weather. This variety is sometimes difficult to establish.

COLIBRI 79 (Meidanover)
Meilland 1981

An improved form of 'Colibri'. The apricot-yellow flowers are veined with orange, and borne on a neat, compact bush. This variety has many flowers open at once, and is ideal for a container, or as a specimen plant.

CORALIN
Dot 1955 Mephisto x Perla de Alcanada

This medium-sized plant bears many flowers in a bright coral red. It is an easily grown variety which adds a vivid touch to any garden.

CORAL ISLAND (Meisevari)
Meilland 1981

The well-shaped blooms of a deep coral salmon are larger than some miniature roses, but the plant is of average size. There are a lot of flowers, and the plant is vigorous and healthy.

COLIBRI

COLIBRI 79 (Meidanover)

CORAL ISLAND (Meisevari)

CREAM GOLD

CORNSILK (Savasilk)

CORNSILK (Savasilk)
Saville 1983 Rise 'n Shine x Sheri Anne

The blooms, normally medium yellow, are inclined to change colour with differing temperatures. In the heat they are very pale, almost white, while in the cool weather they go almost pink.

CRAZY QUILT (Mortrip)
Moore 1980 Little Darling x unnamed miniature seedling

This is another of the striped miniature roses. The flowers have pink and white stripes, and no two flowers have the same markings. The plant is bushy and compact.

CREAM GOLD
Moore 1978 Golden Glow x unknown

The buds are of a classic hybrid tea shape, varying from pale cream to a medium yellow during the season. This plant grows strongly and can become quite large, although judicious pruning will keep it manageable. Do not put this variety in a sheltered part of the garden, as mildew can be a problem.

CRICRI
Meilland 1958 (Alain x Independence) x Perla de Alcanada

The large coral pink flowers have many petals and open flat. The plant is quite large and the buds are globular.

CUDDLES

DEBBIE

CUPCAKE

CUDDLES
Schwartz 1978 Zorina x unnamed seedling

The fully double deep coral pink flowers are petite, and the petals reflex attractively. The compact plant has lots of flowers, usually in trusses, but it can be disbudded for exhibiting.

CUPCAKE
Spies 1981 Gene Boerner x (Gay Princess x Yellow Jewel)

The medium pink flowers are well shaped, high centred and have reflexing petals. The colour holds well both on the bush and when picked. Compact and vigorous, the bush has plenty of foliage. This variety may need protection against rust in some districts.

CUPID'S BEAUTY
Williams 1978 Seedling x Over The Rainbow

The long, pointed buds open to flowers of light orange with a cream reverse. The bush tends to spread and the foliage is sometimes larger than that of most miniatures.

DEBBIE
Moore 1966 Little Darling x Zee

The flowers are yellow with a pink edge, on a bush which can at times send out rather tall shoots.

DEEP VELVET
Jolly 1981 (Grand Opera x Jimmy Greaves) x Baby Katie

This rose with its velvety red blooms and good form is sure to be popular. Even though the flower opens a little quickly it is attractive at all stages. The medium-sized bush has dark green foliage.

DEEP VELVET

DESERT CHARM

DOROLA (Macshana)

DREAMBOAT

DESERT CHARM
Moore 1973 Baccara x Magic Wand

Deep red flowers are well shaped and slightly fragrant. The bush is of medium height, upright, and there is dark leathery foliage. This variety is slower to repeat flower than many in our garden.

DON MARSHALL (Morblack)
Moore 1982 Baccara x Little Chief

We haven't seen this rose but we know it is dark red and the trade name suggests that it is very dark. The bush is spreading with small dark green foliage. The small full-petalled blooms have a slight fragrance.

DOROLA (Macshana)
McGredy 1983 Minuetto x Mabella

This plant seems almost like a floribunda at first glance. It has many unfading yellow blooms of a good shape but both flower and bush size are larger than most miniature roses. Sam McGredy has called this one of the new 'patio' roses, and it would be good grown in a container.

DREAMBOAT
Jolly 1982 Rise 'n Shine x Grand Opera

The very double flowers are deep yellow with the outer petals becoming a little paler as the flower ages. The bush is healthy with small foliage.

DREAMGLO
Williams 1978 Little Darling x
Little Chief

This vigorous bush is a real eye-catcher in the garden. The blooms are silvery white edged with a deep currant red which becomes more predominant as the flowers age. The well-shaped buds open to exhibition type flowers which last well when picked. This rose is good for garden display and for exhibiting.

DRESDEN DOLL
Moore 1975 Fairy Moss x unnamed hybrid moss seedling

This is the most mossed continuous blooming bush type rose we have grown to date. The dusky pink cupped blooms show lots of golden stamens when fully open, and are most attractive. The foliage is medium green and there are no disease problems.

DWARFKING (Zwergkönig)
Kordes 1957 World's Fair x Tom Thumb

The small double flowers are a deep velvety red, usually several to a stem. The plant grows well to a medium size, and the dark green foliage sets off the flowers to perfection.

EASTER MORNING
Moore 1960 Golden Glow x Zee

The very full-petalled larger flowers of ivory white sometimes show a greenish tinge in the cooler weather. Because of the large number of petals, the flowers are inclined to ball in the damp.

DREAMGLO

DRESDEN DOLL

DWARFKING (Zwergkönig)

EASTER MORNING

FAIRLANE

FAIRY MOSS

FAIRLANE
Schwartz 1981 Charlie McCarthy x
unnamed seedling

Elegant high-centred hybrid tea-
shaped flowers are a beautiful
blend of creamy pink and pale
yellow. The flowers pick well, and
some growers say they are
fragrant. The plants are sometimes
inclined to sulk when they are
transplanted and take a while to
establish.

THE FAIRY
Bentall 1932 Lady Godiva sport

This is technically a polyantha, but
with its large clusters of light pink
rosette-shaped small flowers it is
often mistaken for a miniature.
The foliage is larger than a
miniature's, and is pale green. We
find this rose looks very good in a
hanging basket or container, as the
weight of the large clusters of
flowers pull the stems down.

FAIRY MOSS
Moore 1969 (Pinocchio x William
Lobb) x New Penny

This variety was one of the first
miniature moss roses. It has semi-
double mauve pink flowers, mainly
several to a stem. The moss is not
very pronounced. We have found
this variety looks good as a
flowering pot plant.

FIESTA GOLD
Moore 1970 Golden Glow x Magic
Wand

Golden yellow-orange semi-single
flowers cover a strong bush and
make a bright splash of colour in
the garden. The flower shape is at
its best when fully open.

FIESTA GOLD

FIESTA RUBY

FIRE PRINCESS

FUNNY GIRL (Jacfun)

GALAXY (Morgal)

FIESTA RUBY
Moore 1977 Red Pinocchio x Little Chief

This strong-growing plant with large foliage has ruby red well-shaped flowers which may sometimes be lighter in hot weather.

FIRE PRINCESS
Moore 1969 Baccara x Eleanor

The bright red-orange flowers have a globular bud opening to a large flat flower. This was one of the first varieties in this colour range.

FREEGOLD (Macfreego)
McGredy 1981 Seaspray x Dorola

Unfading bronzy yellow flowers of hybrid tea shape on a large upright plant. This is another of the new 'patio' roses.

FUNNY GIRL (Jacfun)
Warriner 1983 Bridal Pink x Fire Princess

The bushy plant has clear pink well-formed blooms which open in abundance. The foliage is dense and dark green.

GALAXY (Morgal)
Moore 1980 Fairy Moss x Fire Princess

This rose has the darkest velvety red flowers of any variety we grow. Because it has many small petals, the flower is not a classic hybrid tea shape, but it is well worth growing for the colour. The foliage is dark green and complements the flowers well.

GOLD COIN

GENEVIEVE (Savagen)
Saville 1983 Unknown miniature climber x unknown miniature seedling

A bedding variety that is covered with small bright yellow flowers. When grown in the sun the edges of the petals are touched with scarlet. The very double blooms form many pointed stars as the flower opens. The plant is upright and bushy with semi-glossy foliage.

GEORGETTE
Bennett 1981 Electron x Little Chief

The petals are a medium rose pink with a deeper pink veining which gives a lacy look. As the flower ages the outer petals lighten in colour, adding to the effect. There are many flowers on a strong-growing bush, with a lot of bright green foliage.

GLORIGLO
Williams 1976 Seedling x Over The Rainbow

An unusual colour combination of glowing orange on the inside of the petals and creamy white on the reverse. The flowers are inclined to have green vegetative centres in the early spring, but lose this tendency as the season progresses.

GOLD COIN
Moore 1967 Golden Glow x Magic Wand

This small variety is good as a rock garden plant, or for use as a flowering plant in a small container. The bright golden yellow flowers are globular in shape and have short stems. You may need to protect against mildew.

GENEVIEVE (Savagen)

GLORIGLO

GOLDEN ANGEL

GOLDEN SONG

GOLD 'N FLAME

GREEN DIAMOND

GOLDEN ANGEL
Moore 1975 Golden Glow x No. 27-62-3 (Little Darling x seedling)

This is a strong-growing plant bearing double medium to deep yellow flowers of a good shape. The flowers are borne singly or several together and pick well.

GOLDEN SONG
Williams 1980 Little Darling x Golden Angel

This climber has golden yellow flowers with deep orange highlights on the petal edges. The long pointed buds open to well-shaped hybrid tea type blooms. The plant grows to about 1.5 to 2 m.

GOLD 'N FLAME
Williams 1980 Unnamed seedling x Over The Rainbow

A bright red and gold bicolour with good colour stability. We have found it rather prone to mildew in our climate.

GREEN DIAMOND
Moore 1975 Unnamed seedling x Sheri Anne

This variety is a real novelty. The dusky pink buds open to white and change to a greenish tinge as the flowers age. The petals are an unusual diamond shape, and the small flowers never open more than about the three-quarter stage. The plant is slower to repeat flower than some varieties.

GREEN ICE
Moore 1971 (*R. wichuraiana* x Floradora) x Jet Trail

Globular pale pink buds open to rosette-shaped flowers of white which will turn to pale green as they age. The greenish colour will be more definite if grown in semi-shade. There are many flowers on each stem and the plant looks very good in a container or hanging basket.

GYPSY JEWEL
Moore 1975 Little Darling x Little Buckaroo

The flowers are a deep rose pink, with the buds a lighter colour on the reverse. Flowers, leaves and plant are all larger than usual.

HEARTLAND (Savsay)
Saville 1982 Sheri Anne x Watercolour

The beautiful soft rich coral flowers last well. The form of the blooms is rather variable, and the flower is large. The plant is vigorous and upright and with its abundant bloom makes a good bedding rose.

HEIDI
Christensen 1978 Fairy Moss x Iceberg

The very double flowers are a clear medium pink and have a lot of moss on the sepals. It is a very vigorous grower and repeat blooms quickly.

GREEN ICE   HOKEY POKEY (Savapok)

GYPSY JEWEL

HEIDI

HI-DE-HI

HOMBRE

HONEST ABE

HI-DE-HI
McGredy 1981 Anytime x Otto Linne

Mauve pink flowers with narrow petals form very unusually shaped flowers.

HIGH SPIRITS (Savaspir)
Saville 1984 Sheri Anne x Tamango

The medium red well-shaped flowers come in large sprays on a vigorous, tall and upright plant.

HOKEY POKEY (Savapok)
Saville 1980 Rise 'n Shine x Sheri Anne

This delightful rose is a unique colour somewhere between a deep apricot and orange with yellow undertones. The flowers are semi-double, and sometimes lose colour in the heat. The vigorous well-branched plant is easy to grow.

HOMBRE
Jolly 1982 Humdinger x Rise 'n Shine

The small well-formed blooms have reflexing petals shading from light pink on the outside to creamy apricot in the centre. The plant is bushy and rather short.

HONEST ABE
Christensen 1978 Fairy Moss x Rubinette

The very mossy buds open to deep velvety crimson red double flowers. The plant is vigorous and bushy, but will get some mildew in sheltered situations.

HONEYCOMB
Moore 1974 (*R. wichuraiana* x
Floradora) x Debbie

The flowers are a soft buff yellow
with many petals. The foliage is
small and glossy and the plant is
dwarf and bushy.

HULA GIRL
Williams 1975 Miss Hillcrest x
Mabel Dot

The well-shaped buds and double
flowers are a bright orange with a
touch of yellow at the base of the
petals. There is a fruity fragrance,
particularly in the heat. Bushy and
well branched, the plant is
moderately compact.

HUMDINGER
Schwartz 1976 Gold Coin seedling
x unnamed miniature seedling

The very double flowers are a coral
orange, on a vigorous upright
plant. There is profuse bloom
which repeats quickly.

HONEYCOMB

HULA GIRL

HUMDINGER

JANICE

HUMPTY DUMPTY

HUMPTY DUMPTY
De Vink 1952 (*R. multiflora nana* x Mrs Pierre S. du Pont) x Tom Thumb

The small very double flowers are a soft carmine pink, borne in clusters. An older variety which is not often offered for sale.

JANICE
Moore 1971 (*R. wichuraiana* x Floradora) x Eleanor

Globular buds open to small double flowers of deep rose pink, several to a stem. The foliage on the vigorous upright bush is dark green and glossy.

JANNA
Moore 1970 Little Darling x (Little Darling x (*R. wichuraiana* x miniature seedling))

Flowers of rose pink with white reverse have a good bud shape. The leathery foliage completes a healthy, bushy plant.

JANNA

JEANIE WILLIAMS
Moore 1965 Little Darling x Magic Wand

Small hybrid tea-shaped flowers of yellow overlaid with scarlet are in proportion to the short bushy plant.

JEANNE LAJOIE
Sima 1975 (Casa Blanca x Independence) x Midget

This vigorous branching climber has well-formed pointed buds and double flowers of medium pink with slightly darker reverse. This is a true climber which needs training on a fence or trellis.

JELLY BEAN
Saville 1982 Unnamed seedling x Poker Chip

This little brilliant red rose opens to show a yellow eye. The bush is compact, bushy and vigorous and grows to only 25 cm.

JET TRAIL
Moore 1964 Little Darling x Magic Wand

This variety has been around for a while now, but is still one of the better whites. The pointed buds open to double flowers which last well on the bush. The stamens darken almost to grey as the flower ages.

JOSEPHINE WHEATCROFT
(Rosina)
Dot 1951 Eduardo Toda x Rouletii

The sunflower yellow semi-double flowers open quickly. They make a good splash of colour in the garden, and contrast well with the glossy foliage. This is another variety which will need protection from mildew.

JEANIE WILLIAMS

JEANNE LAJOIE

JELLY BEAN

JUDY FISCHER

JOSEPHINE WHEATCROFT (Rosina)

JULIE ANN (Savaweek)

KATHY ROBINSON

KAIKOURA (Macwalla)

JUDY FISCHER
Moore 1968 Little Darling x Magic Wand

The double flowers are a deep rose pink which may be lighter in hot weather. The plant grows vigorously, and is easy to establish. The flowers last well on the bush, and when cut.

JULIE ANN (Savaweek)
Saville 1984 Zorina x Poker Chip

Hybrid tea-shaped buds and flowers of a brilliant vermilion orange come in profusion on a healthy bush throughout the season.

JUNE TIME
Moore 1963 (*R. wichuraiana* x Floradora) x ((Étoile Luisante seedling x Red Ripples) x Zee)

The double light pink flowers are borne singly and several together on a moderately compact plant.

KAIKOURA (Macwalla)
McGredy 1978 Anytime x Matangi

The larger flowers of bright orange-red open flat. The foliage is dark green and glossy and the plant is vigorous, bushy and free blooming.

KATHY ROBINSON
Williams 1968 Little Darling x Over The Rainbow

This is one of those in-between varieties — not quite a climber, but a tall grower. The well-shaped flowers of deep pink with a cream-buff reverse are on long stems, making them good for picking. Judicious cutting back during the flowering season will keep this rose down to a manageable size.

KO'S YELLOW (Mackosyel)

KOA
Rovinski/Meredith 1978 Persian Princess x Gene Boerner

Double coral pink flowers with a hybrid tea-shaped bud on a moderately tall, open bush.

KO'S YELLOW (Mackosyel)
McGredy 1978 (New Penny x Banbridge) x (Border Flame x Manx Queen)

The larger flowers are a classic hybrid tea shape. They are medium yellow, sometimes with a deep pink edge on the petals. The glossy foliage is also larger, but in proportion to the flower size. We find this variety to be very fragrant in warm weather.

LADY EVE
Rovinski/Meredith 1978 Neue Revue x Sheri Anne

The round bud and double flowers are cream with a coral edge, on a tall spreading bush.

LAVENDER JEWEL
Moore 1978 Little Chief x Angel Face

The soft lavender buds open to exhibition form flowers, each petal edged with a thin magenta margin. The dark foliage tones well.

LAVENDER LACE
Moore 1968 Ellen Poulsen x Debbie

The many-petalled flowers are a soft lavender pink and open rather quickly to a loose form. The plant is short and slow growing, and the individual flowers are on short stems.

LAVENDER JEWEL

LELIA LAIRD

LAVENDER LACE

LELIA LAIRD

Bennett 1979 Contempo x Sheri Anne

The coral flowers have a touch of yellow blending and a prominent yellow eye when fully open. Both flowers and plant are a little larger than average. Blooms, which are often one to a stem, repeat quickly.

LEMON DELIGHT

Moore 1978 Fairy Moss x Goldmoss

This unusual moss rose has a distinct citrus fragrance when the oil glands in the moss are gently touched. The delightful lemon buds open to ten-petalled blooms with prominent stamens and the plant grows and flowers well.

LIBBY

Rovinski/Meredith 1978 Overture x Perla de Alcanada

The semi-double flowers are white with a red edge on a compact, moderately vigorous bush.

LITTLE BUCKAROO

Moore 1956 (*R. wichuraiana* x Floradora) x (Oakington Ruby x Floradora)

This variety gives a good display in the garden, with many bright scarlet red blooms open at the same time. The bush is taller than average with dark glossy foliage.

LEMON DELIGHT

LIBBY

LITTLE BUCKAROO

LITTLE GIRL

LITTLE DARLING
Duehrsen 1956 Capt. Thomas x
(Baby Chateau x Fashion)

Definitely not a miniature but
worth including as this floribunda
is found in the background of many
of our present day miniature roses.
It sets seed readily and its
influence is clearly seen in the
flower shape of varieties such as
'Beauty Secret', 'Mary Marshall',
'Sheri Anne' and 'Rise 'n Shine'.

LITTLE FLIRT
Moore 1961 (*R. wichuraiana* x
Floradora) x (Golden Glow x Zee)

Rather loose flowers of red and
yellow are borne on a medium-sized
bush, making a bright splash of
colour in the garden.

LITTLE GIRL
Moore 1973 Little Darling x
Westmont

'Little Girl' grows to be a big girl —
usually 2 to 2.5 m tall. This climber
has shapely coral pink blooms
borne in profusion. The plant is
upright, like a pillar rose, and
needs some staking for the first
season or two. The foliage is dark
and glossy and the plant is
practically thornless.

LITTLE DARLING

LITTLE FLIRT

LITTLE JACKIE (Savor)
Saville 1982 (Prominent x Sheri
Anne) x Glenfiddich

The delicate pastel flowers of an
orange, pink and yellow blend are a
beautiful hybrid tea shape, and the
petals reflex well. The plant is
vigorous and well branched.

LITTLE LINDA
Schwartz 1976 Gold Coin seedling
x unknown miniature seedling

This short compact plant has many
small light lemon flowers, which
may be tipped with pink in the
cooler weather.

LITTLE RASCAL
Jolly 1981 Sheri Anne x Rise 'n
Shine

The flower is red, shading to
yellow at the base of the petals,
with a well-shaped bud. The
opening flower is of exhibition type.
We have experienced some
difficulty with mildew when this
rose is grown in a greenhouse, and
have read reports about a slight
lack of winter hardiness.

LITTLE SCOTCH
Moore 1958 Golden Glow x Zee

The buff-cream buds open to
informal near white flowers on a
vigorous open plant. The colour
will be more stable if grown in
dappled shade.

LITTLE LINDA

LITTLE JACKIE (Savor)

LITTLE RASCAL

LITTLE SCOTCH

LITTLE SHOWOFF
Moore 1960 Golden Glow x Zee

Another climber, this time in a golden yellow with orange overtones. The flowers are semi-single and a loose shape. Like others, it will need training on a fence or trellis.

LITTLE SHRIMP
Lens 1983

The coral pink well-shaped flowers are larger than usual, with the foliage and bush in proportion.

LITTLE SHRIMP

LITTLE SHOWOFF

LOVING TOUCH
Jolly 1982 Rise 'n Shine x First
Prize

This is a most welcome addition to
the true apricot colour range. The
flowers are a very good shape,
especially at the bud stage. The
foliage and flowers are both a little
larger than usual and the bush is
inclined to spread.

LYNN ANN
Saville 1981 Rise 'n Shine x Sheri
Anne

The orange-yellow flowers have
many petals and are usually borne
in sprays. The plant is well
branched, with many flowers open
at any one time.

MACSPICE (Macspike)
McGredy 1983 Anytime x Otto
Linne

The long flowering spikes are
tipped with many small mauve
blooms. The plant has small,
medium green foliage, and the
growth is spreading. This variety
is sometimes classified as a shrub
rose.

MADELYN LANG
Williams 1970 Little Darling x
Little Chief

The many small deep pink to red
flowers grow in clusters on a tall
plant which can be cut back, or
used as a short climber.

LOVING TOUCH

LYNN ANN

MADELYN LANG

MAGIC CARROUSEL

MAGIC CARROUSEL
Moore 1972 Little Darling x
Westmont

The striking blooms of white edged
with red are borne in profusion on
a tall vigorous plant. It is typical of
this variety to throw up strong new
shoots with many flowers. When
the flowers have faded, cut back to
about half its height to keep the
plant more manageable.

MAGIC DRAGON
Moore 1969 ((*R. wichuraiana* x
Floradora) x seedling) x Little
Buckaroo

This is the only dark red miniature
climber available at present. The
loose flowers usually come in
clusters and are slower to repeat
than many varieties.

MAIDY (Korwalbe)
Kordes 1984 Regensberg x seedling

We haven't seen this rose. The
description supplied with the
photograph reads, 'Elegant buds
like a hybrid tea, colour dark red
with white reverse, compact low
growing, very vigorous.'

MAGIC DRAGON

MAIDY (Korwalbe)

MAORI DOLL
Bell 1977 Yellow Doll sport

The buff yellow flowers of this variety keep their colour better if planted in semi-shade. The plant grows in a similar way to 'Yellow Doll'.

MARILYN
Dot 1955 Perla de Montserrat x Bambino

Small pale pink rosette-like flowers smother the compact bush, and make a good show in the garden. Like many of the older varieties it does mildew.

MARK ONE (Savamark)
Saville 1982 Sheri Anne x Glenfiddich

The brilliant orange flowers are borne in abundance and set off by the dark green glossy foliage. It is at its best when freshly open, becoming more pink as it ages. The plant is bushy and almost constantly in bloom.

MARY ADAIR

Moore 1966 Golden Glow x Zee
The well-shaped flowers are a true soft apricot. The moderately short plant is easy to grow and gives many flowers suitable either for garden display or for picking.

MARY MARSHALL
Moore 1970 Little Darling x Fairy Princess

This is the rose we recommend to people who have never grown a miniature before. It seems to have no vices, only virtues. The coral orange flowers have a yellow base to the petals; they have a good shape and repeat quickly. The plant grows well and throws up many basal shoots. It appears to be almost disease resistant.

MAORI DOLL

MARK ONE (Savamark)

MARILYN

MARY ADAIR

MARY MARSHALL

MAX COLWELL

MEILLANDINA (Meirov)

MIDGET

MIMI

MAX COLWELL
Moore 1969 Unnamed floribunda seedling x (Little Darling x miniature seedling)

This is another variety which cannot decide whether it is a low climber or a tall bush, so it can be treated as either. The double red flowers often come one to a stem, and last well when picked.

MEILLANDINA (Meirov)
Meilland 1975 Rumba x (Dany Robin x Fire King)

The vigorous bushy plant has many loose flowers of a scarlet red with a yellow centre. It makes a good pot or container plant. This is one of a series of varieties using the name 'Meillandina' — 'Orange Meillandina', 'Scarlet Meillandina' and, we understand, 'Yellow Meillandina' and 'Apricot Meillandina'. Not all of them have been raised by Meilland. Harmon Saville tells us that 'Apricot Meillandina' is actually 'Savamark', the miniature sold in the United States as 'Mark One'. Each one is reputed to be one of the best in its colour for use as a bedding miniature.

MIDGET
De Vink 1941 Ellen Poulsen x Tom Thumb

Small light carmine red flowers with a paler centre are borne in large clusters on a moderately compact plant with fine foliage.

MIMI
Meilland 1965 Moulin Rouge x (Fashion x Perla de Montserrat) A mature plant of 'Mimi' can reach 1 m high and across, with many flat rosette flowers in a medium pink, and light green foliage.

MINIJET

We have not been able to find any information about the raiser or breeding of this rose which has been on sale in New Zealand for a number of years. It has many soft pink flowers, often borne in large clusters, on a compact plant. We personally have found it to be rather ordinary in the garden, but it comes into its own when grown in the greenhouse, where this photograph was taken.

MINNIE PEARL (Savahowdy)
Saville 1982 (Little Darling x Tiki) x Party Girl

The long elegant buds open to well-shaped blooms of pink with a touch of yellow. The colour deepens in the hot sun. Blooms come one to a stem on an upright, well-shaped bush.

MINUETTO (Darling Flame) (Meilucca)
Meilland 1971 (Rimona x Josephine Wheatcroft) x Zambra

The orange-red buds open to brilliant orange-yellow semi-double larger flowers. This is another variety which is good in a pot.

MINIJET

MINNIE PEARL (Savahowdy)

MINUETTO (Darling Flame (Meilucca)

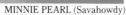

MISTY DAWN

MOANA (Macbipi)

MOOD MUSIC

MR BLUEBIRD

MY VALENTINE

MISTY DAWN
Schwartz 1979 Charlie McCarthy x unnamed seedling

The pure white flat flowers have many small petals, and contrast very well with the fine dark green foliage. This variety will often send up vigorous basal shoots which should be cut back to half their length when the flowers have faded.

MOANA (Macbipi)
McGredy 1978 Seedling x New Penny

Dark green foliage complements the medium to light pink flowers which show yellow stamens when fully open. This is a vigorous showy variety in the garden.

MOOD MUSIC
Moore 1977 Fairy Moss x Goldmoss

The globular buds are very mossy, and open to double peach-orange flowers. The plant is moderately compact and very thorny.

MR BLUEBIRD
Moore 1960 Old Blush x Old Blush

It is unfortunate that the flowers of this variety are of such a loose form and drop so quickly when picked, as it is one of the deepest mauve-lavenders. The plant has quite coarse foliage and can grow rather tall.

MY VALENTINE
Moore 1975 Little Chief x Little Curt

We were told that this variety was bred specifically for the Valentine's Day pot plant trade in the United States, and it certainly is suitable. The flowers — dark red for love — are borne in clusters above bronze-green foliage. The compact plant looks good in a pot. With encouragement, it will flower under lights out of season.

NANCY HALL
Moore 1972 Mary Adair sport

We have grown this variety for several years, and find it almost indistinguishable from 'Mary Adair', from which it is a sport. The main difference seems to be a greater propensity to mildew.

NEW PENNY
Moore 1962 (*R. wichuraiana* x Floradora) x unnamed seedling

The semi-double orange-pink flowers come in clusters on the end of strong tall shoots which may need staking if growing in a windy area. Cutting these shoots back after flowering will help to keep the plant compact.

NOZOMI
Onodera 1968 Fairy Princess x Sweet Fairy

This is a rose which almost defies description. When we have grown it as a groundcover, we have had shoots reach 2 to 2.5 m in a season. The pale pink buds open to small five-petalled flowers which lighten as they age and come more prolifically on second-year wood. Prune very lightly, if at all. This variety also looks good in a hanging basket. Unfortunately it flowers mainly in the spring, with only a few flowers during the rest of the season.

NANCY HALL

NEW PENNY

NOZOMI

OLYMPIC GOLD

OLYMPIC GOLD
Jolly 1982 Rise 'n Shine x Bonny

We have watched this plant closely for two seasons, and have doubts as to whether it can be considered a miniature under our growing conditions. The flower is well shaped but almost as large as a floribunda, and the deep yellow of the bud fades to cream by the time the flower is fully open. The foliage and bush are also large, in proportion to the size of the flower.

ORANGE FIRE
Moore 1974 (*R. wichuraiana* x Floradora) x Fire Princess

These flowers are more pink than orange for us, on an easily grown plant. We have read that this variety will do better with extra iron.

ORANGE HONEY
Moore 1979 Rumba x Over The Rainbow

The yellow-amber buds open to brilliant orange flowers with a yellow reverse, changing to red as they age, giving an autumn colour effect. It is a most eye-catching plant in the garden, and can also be very effective in a hanging basket.

ORANGE MEILLANDINA
(Meijikatar)
Meilland 1982 Meichanso x (Meidacinu x Duchess of Windsor)

This variety is also sold as 'Sunblaze' and 'Orange Sunblaze'. The large flowers are an orange vermilion, on an upright bushy plant with dull light green foliage. Good for garden display or for growing in a container.

ORANGE HONEY

ORANGE PIXIE
Moore 1978 Little Chief x Fire Princess

The deep orange-red flowers are quite small, as is the plant. We have been told that this variety does not like being transplanted, and will sometimes sulk for a season before getting established.

ORANGE TIME
Bell 1983

Deep orange blooms are borne in clusters on a neat compact bush.

OTAGO (Macnecta)
McGredy 1978 Anytime x Minuette

The medium-sized double salmon red flowers come in profusion on a bushy plant.

OVER THE RAINBOW
Moore 1972 Little Darling x Westmont

The well-shaped flowers are deep scarlet red with a bright yellow reverse. The bush is vigorous and will grow quite tall. It will get mildew if you neglect to spray it. Unfortunately, this tendency is also passed on to its offspring.

ORANGE PIXIE

OVER THE RAINBOW

ORANGE TIME

OTAGO (Macnecta)

PACESETTER

PACESETTER
Schwartz 1979 Ma Perkins x Magic Carrousel

From the first time we saw this rose we have loved it. The pure white flowers are a perfect hybrid tea shape in miniature, and can be easily disbudded to give long stems for picking. The bush can grow quite tall, but this means more flowers, which can only be a bonus.

PAINTBRUSH
Moore 1975 Fairy Moss x Goldmoss

The long soft apricot buds would be attractive even if they were not extremely mossy. Unfortunately the semi-single flowers fade almost to white. The plant is vigorous and very thorny.

PAINTBRUSH

PARAGON (Mackosred)

PEACHES 'N CREAM

PARTY GIRL (Savapart)

PEACHY WHITE

PEACHY KEEN

PARAGON (Mackosred)
McGredy 1982 Ko's Yellow x Little Artist

The deep rose pink to red flowers are a little larger but have a good shape and pick well. The plant is very vigorous, and will grow quite tall.

PARTY GIRL (Savapart)
Saville 1979 Rise 'n Shine x Sheri Anne

Although the flowers have fewer petals than some varieties, they last well both on the bush and when picked. They are a beautiful blend of soft yellow flushed with salmon pink. The upright vigorous plant has a tendency to mildew if grown in a sheltered spot.

PEACHES 'N CREAM
Woolcock 1976 Little Darling x Magic Wand

This is a very well-named variety, with the delicious blend of peachy pink and cream looking almost good enough to eat. The flowers are a good shape, but have so many petals that they sometimes do not open completely in cold or damp weather. There are many flowers on a compact plant.

PEACHY KEEN
Bennett 1979 Little Darling x Sheri Anne

The long pointed buds are a soft apricot pink and cream blend. Although there are fewer petals than some varieties, the flowers hold well at the half-open stage with a very crisp petal texture, and last well when picked. There seem to be very few thorns.

PEACHY WHITE
Moore 1976 Little Darling x Red
Germain

Although this has the same
number of petals as the previous
variety, it goes very quickly from
bud to fully open bloom. But the
buds are very appealing, being
nearly white with a delicate edge of
peach pink and the fully open
bloom lasts well.

PEEK A BOO (Dicgrow) (Brass
Ring)
Dickson 1981 (Bangor x Korbell) x
Nozomi

The well-shaped soft salmon pink
buds open into loose flowers which
lose colour quickly in the heat. The
plant can grow quite large and
leggy.

PENNY CANDY
Saville 1981 Rise 'n Shine x Sheri
Anne

The plants of this variety which
we grow are small with very fine
foliage and very small orange-
yellow flowers. They seem to get
mildew very easily.

PEON (Tom Thumb)
De Vink 1936 Rouletii x Gloria
Mundi

The first commercial seedling from
'Rouletii'. The flowers are reported
to be small and dark red with a
white centre. The plant is very
dwarf and very free blooming.
Although this variety is not often
grown or seen today, it is very
important in the background of
miniature roses.

PEEK A BOO (Dicgrow) (Brass Ring)

PENNY CANDY

PERLA DE MONTSERRAT

PERLA DE ALCANADA
Dot 1944 Perle des Rouges x Rouletii

This variety can be obtained both as a bush and a climber. The latter is more common in New Zealand. The small, semi-double carmine red blooms are borne mainly on second-year wood on the climbing version. The climbing plant will make a dense thicket if left without pruning.

PERLA DE MONTSERRAT
Dot 1945 Cécile Brunner x Rouletii

The semi-double pale pink rosette-like flowers are on a small compact plant. Because there are many flowers open at once on a many-branched plant, it looks good in a container, or as a specimen in a rock garden.

PETITE FOLIE
Meilland 1968 (Dany Robin x Fire King) x (Cricri x Perla de Montserrat)

The bright coral orange flowers are a good shape and come in profusion on an excellent strong-growing plant.

PETTICOAT
Warriner 1981 Bon Bon x Lemon Delight

The many-petalled white blooms come in clusters and open flat. The plant is free blooming all through the growing season. Although the growth is rather spreading, the plant is still quite compact.

PETITE FOLIE

PETTICOAT

PINK CAMEO
Moore 1954 (Soeur Thérèse x
Skyrocket) x Zee

The soft pink flowers are borne in
profusion on a vigorous climbing
plant which can be trained into a
typical fan shape. The foliage is a
light green.

PINK CASCADE (Morcade)
Moore 1981 (*R. wichuraiana* x
Floradora) x Magic Dragon

The rosette flowers are quite a
deep pink on a plant similar in
growth to 'Red Cascade'.

PINK CHERUB (Morfair)
Moore 1980 Fairy Moss x Fairy
Moss

The medium to light pink flowers
are often lighter in colour at the
tips of the petals. The flowers are
double and come in small clusters
on a very bushy plant.

PINK CASCADE (Morcade)

PINK CAMEO

PINK CHERUB (Morfair)

PINK DELIGHT (Lenpi)

PINK DELIGHT (Lenpi)
Lens 1982

The beautiful pastel pink larger flowers are a joy from the bud right through to the fully open stage. The bush grows well, except for a slight tendency towards mildew.

PINK JOY
Moore 1953 Oakington Ruby x Oakington Ruby

The loose flowers are a deep pink, with a globular bud. The plant is moderately compact.

PINK MANDY
Moore 1974 Ellen Poulsen x Little Chief

Very double deep rose pink flowers come several together on each stem, giving the impression of a miniature floribunda. Strong-growing basal shoots will grow quite tall and should be cut back after the flowers have finished.

PINK JOY

PINK MANDY

PINK PETTICOAT
Strawn 1979 Neue Revue x Sheri Anne

The well-shaped flowers of creamy white brushed with deep coral pink along the outer edges often come one to a stem, and are larger than usual. The bush is quite tall, vigorous and healthy. Beautiful in the garden and for picking.

PINK TRIUMPH
Jolly 1983 Operetta x Bonny

The clear salmon pink blooms are a good shape, and have a definite fragrance. The plant is upright and bushy and the flowers are good for picking.

PIXIE
De Vink 1940 Ellen Poulsen x Tom Thumb

The very double white flowers are small, with a globular bud. The foliage is fine and the bush small and compact.

PLUM DUFFY
Bennett 1978 Magic Carrousel x Magic Carrousel

The well-shaped buds and flowers are an unusual colour ranging from plum to burgundy, on a plant which is sometimes slow to get established.

PINK PETTICOAT

PLUM DUFFY

POKER CHIP

PINK TRIUMPH

POKER CHIP
Saville 1979 Sheri Anne x (Yellow Jewel x Tamango seedling)

Although the eye-catching blooms of scarlet-orange with golden reverse are a little large, they have an excellent shape, and are a focal point of colour in any garden. The plant grows well, but never gets out of control.

POPCORN
Morey 1973 Katharina Zeimet x Diamond Jewel

Clouds of tiny semi-single flowers with yellow stamens are borne in clusters on a vigorous plant with fine, almost fern-like foliage. We have found 'Popcorn' to be very fragrant in warm weather.

POUR TOI (Para Ti)
Dot 1946 Eduardo Toda x Pompom de Paris

The larger flowers are semi-single, and white with bright yellow stamens. The dark green foliage is larger, in proportion to the flower size.

POPCORN

PUPPY LOVE (Savapup)

PUPPY LOVE (Savapup)
Schwartz 1978 Zorina x unnamed
seedling

A flower of excellent shape, and a
beautiful blend of orange, pink and
yellow on a neat compact bush. We
find this one of the earliest to
flower in the spring.

RAGTIME (Maccourlod)
McGredy 1983 Mary Sumner x
unnamed seedling

One of the first 'hand-painted'
miniatures, with all flowers
showing different markings. They
are mostly in bright reds and
pinks, some being somewhat paler.
The foliage is dark and the bush
compact and vigorous.

RAINBOW'S END (Savalife)
Saville 1984 Rise 'n Shine x
Watercolour

These hybrid tea-shaped flowers of
golden yellow with scarlet-edged
petals will be good for exhibitors as
well as in the garden. The flowers
will be a pure deep lemon yellow
when grown in shade or under
cover. The plant is vigorous and
compact.

RED ACE
Saville 1980 Rise 'n Shine x Sheri
Anne

The free-blooming currant red
flowers have good form but open
rather quickly. The small bushy
upright plant has dark green
foliage. It is easy to grow and
makes a good container plant.

RAGTIME (Maccourlod)

RED BEAUTY

RED ACE

RAINBOW'S END (Savalife)

RED CASCADE

RED FLUSH

RED ACE (Amruda)
De Ruiter 1982

There seem to be two roses with this name. Look for the variety name (Amruda) or the breeder, to distinguish between them. This miniature is a deep crimson of good form. It flowers in profusion on a compact disease-resistant bush.

RED BEAUTY
Williams 1981 Starburst x Over The Rainbow

The dark red blooms usually come one to a stem, and are eye catching in the garden. The plant is bushy and symmetrical with dark green foliage.

RED CASCADE
Moore 1976 (*R. wichuraiana* x Floradora) x Magic Dragon

This unusual rose is a prostrate or cascading climber. As a groundcover it can grow 2 m across with the canes lying on the ground. At first the dark red flowers come at the end of the stems, then laterals grow at every leaf, each with its cluster of flowers. It really comes into its own draped over a wall or down a bank, or in a hanging basket. Some protection against mildew will be needed.

RED FLUSH
Schwartz 1978 No parentage given

The double medium red blooms come in clusters on a healthy profusely blooming plant which is bright and eye catching in the garden.

RED IMP
De Vink 1951 Ellen Poulsen x Tom Thumb

The globular buds open into small red flowers on a short compact bush. We find that, in our climate, the buds often go brown without opening fully but, grown in a container in the greenhouse, the plant can be quite successful.

RED SHADOWS (Savmore)
Saville 1983

The petals of the crimson red flowers have a darker edge, particularly as the flower ages. There are many flowers on a healthy compact bush, making a good show in the garden. A good dark red rose for the garden.

RED WAGON (Mordan)
Moore 1980 Little Darling x Little Chief

This plant gives a grand splash of colour in the garden. It has many semi-double bright scarlet red blooms on a healthy, well-branched plant.

RISE 'N SHINE
Moore 1977 Little Darling x Yellow Magic

This brilliant clear yellow rose has many exhibition form blooms on a superior bush. The blooms are comparatively unfading, and repeat quickly. This is still the yellow miniature with which all new yellow miniatures are compared.

ROSE HILLS RED
Moore 1978 (*R. wichuraiana* x Floradora) x Westmont

These deep red blooms are at their best when fully open, when they make a good display in the garden or in a container. The plant is vigorous and healthy.

RED IMP

RED SHADOWS (Savmore)

RED WAGON (Mordan)

RISE 'N SHINE

ROSE HILLS RED

ROULETII

ROULETII

One of the original miniature roses. The loose flowers of a lilac pink are borne on a small plant which is sometimes difficult to grow. When you see this variety you realise how much progress modern hybridisers have made with miniature roses.

ROYAL RUBY

Morey 1972 Garnette x (Tom Thumb x Ruby Jewel)

The double ruby red flowers have a good shape and repeat quickly. The plant grows well and is of medium height.

SASSY LASSY

Williams 1975 Seedling x Over The Rainbow

This plant makes a vivid splash of colour in the garden. The flowers are a yellow blend with pink, orange and red shadings. The bushy plant has bronze-green foliage and may need protection from mildew.

SCARLET GEM

Meilland 1961 (Moulin Rouge x Fashion) x (Perla de Montserrat x Perla de Alcanada)

The globular buds open into small scarlet red flowers on a compact bushy plant.

SCARLET MEILLANDINA

Meilland 1980 Meidanu x (Meidacinu x Duchess of Windsor)

The bright scarlet red flowers are larger than usual, and are at their best when fully open, when they show bright yellow stamens. The plant is vigorous and has healthy dark green foliage.

ROYAL RUBY

SCARLET MEILLANDINA

SEABREEZE
Lemrow 1976 Unknown x White Fairy

The pale lilac pink flowers are borne in floribunda-like clusters on a healthy plant which has large pale green foliage.

SELWYN TOOGOOD
Eagle 1983 Heidi seedling

The lightly mossed buds open into small well-shaped flowers of medium rose pink. The foliage is light green and the plant small and slow growing.

SEPTEMBER DAYS
Saville 1976 Rise 'n Shine x Yellow Jewel

Globular buds and loose flowers cover the short plant, making a small mound of yellow in the garden.

SHAKESPEARE FESTIVAL
Moore 1979 Golden Angel x Golden Angel

Larger flowers of a clear golden yellow are borne on a healthy plant which is attractive in the garden.

SEABREEZE

SELWYN TOOGOOD

SHAKESPEARE FESTIVAL

SHERI ANNE

SI

SHOOTING STAR

SIERRA SUNRISE (Morliyel)

SHERI ANNE
Moore 1973 Little Darling x New Penny

Beautifully shaped hybrid tea type buds open to semi-single flowers of a vivid orange-red. The plant is moderately vigorous and compact. It is interesting to note how many times 'Sheri Anne' has been used in the breeding of the newer varieties. It is obviously a good parent.

SHOOTING STAR
Meilland 1972 Rumba x (Dany Robin x Perla de Montserrat)

Brightly coloured orange-yellow blend flowers on a large plant are eye catching in the garden. We find this variety has a tendency to mildew, particularly in the autumn.

SI
Dot 1957 Perla de Montserrat x (Anny x Tom Thumb)

As far as we know, this is the smallest miniature rose in commercial production. The tiny pink buds are no larger than a grain of wheat, and open to semi-double light pink flowers about 1 cm across. The fine foliage and short plant are on a scale to match the flowers.

SIERRA SUNRISE (Morliyel)
Moore 1980 Little Darling x Yellow Magic

The well-shaped flowers are a medium yellow with a pink edge to the petals which becomes more pronounced as the flower ages. The bush is strong growing and healthy. A good variety for picking.

SILVER TIPS

SILVER TIPS
Moore 1961 (*R. wichuraiana* x
Floradora) x Lilac Time

The lilac pink flowers have many
narrow petals, making an unusual
shape. The plant sends up long
shoots with many flowers at the
end. This variety is sometimes
used in hanging baskets.

SIMPLEX
Moore 1961 (*R. wichuraiana* x
Floradora) x unnamed seedling

Apricot buds open to charming
creamy white five-petalled flowers
with yellow stamens. The blooms
come in large sprays at the end of
long stems. Vigorous and easy to
grow, the plant is outstanding in
its simple beauty.

SIMPLEX

SNOW CARPET (Maccarpe)
McGredy 1980 New Penny x
Temple Bells

This groundcover rose forms a
mound of informal creamy white
flowers in spring. Sometimes this
variety will sulk when first
transplanted, and will take a while
to get going.

SNOW WHITE
Robinson 1955 Little Princess x
Baby Bunting

The small white double flowers are
sometimes marked with pale pink
after rain. The foliage is fine and
the plant small.

SNOW WHITE

STACEY SUE

SPICE DROP (Savaswet)

STARGLO

STARINA

SPICE DROP (Savaswet)
Saville 1982 (Sheri Anne x Glenfiddich) x (unnamed moss seedling x [Sarabande x Little Chief])

This small-growing variety has all the desirable features of a miniature rose. The well-formed salmon pink buds and flowers cover the dense well-branched plant which is vigorous and symmetrical with lots of dark green foliage. It is very prolific and almost constantly in bloom.

STACEY SUE
Moore 1976 Ellen Poulsen x Fairy Princess

Small rosette-like blooms of soft pink come in clusters on a short very bushy plant. This variety smothers itself with flowers and repeats quickly.

STARGLO
Williams 1973 Little Darling x Jet Trail

This outstanding white exhibition form bloom has just enough creamy yellow in the centre of the flower to make the buds and flowers glow. When the plants of this variety are small the stems are too thin to hold the large flowers erect, but as the plant matures the stems become much stronger.

STARINA
Meilland 1965 (Dany Robin x Fire King) x Perla de Montserrat

Well-shaped flowers of orange-red are borne singly or several together on a healthy vigorous plant. For many years the most popular miniature rose in England, the United States and New Zealand.

STARS 'N STRIPES

STRAWBERRY SWIRL

STARS 'N STRIPES
Moore 1976 Little Chief x unnamed
seedling

Every flower of this variety is
unique, the red and white striping
differing on every one. This is one
variety that will repeat better and
stay a more compact plant if cut
back after the flowers have
finished.

STRAWBERRY SWIRL
Moore 1978 Little Darling x
unnamed miniature seedling

An unusual blend of red, pink and
white stripes on the blooms makes
this plant well worth growing. The
buds are slightly mossed. Long
arching stems grow during the
season with flowers at the end, and
also on laterals along the stems.
Watch out for the thorns.

SUGAR ELF
Moore 1974 (*R. wichuraiana* x
Floradora) x Debbie

Although this variety is sometimes
called a climber because of its long
arching branches, it looks best in a
hanging basket or a container. The
small flowers are a pink and gold
blend, which contrasts well with
the dark green glossy foliage.

SUGAR ELF

SUMMER BUTTER

SUMMER BUTTER
Saville 1979 Arthur Bell x Yellow
Jewel

The semi-double flowers of medium
yellow are quite fragrant and
repeat quickly. The plant is
reasonably short, but grows
vigorously.

SUNBLAZE
See 'Orange Meillandina'.

SUNDUST

SUNMAID

SUNNYSIDE

SWEDISH DOLL

SUNDUST
Moore 1977 Golden Glow x Magic Wand

The soft apricot yellow buds open to buff yellow semi-double flowers which are fragrant. The colour will be better if the plant is grown in partial shade.

SUNMAID
Spek

This plant is a real attention-getter in the spring. The bright golden orange flowers deepen in full sun. The bush can get quite large, particularly if it is pruned only lightly in the winter. You may need to watch this variety for mildew and rust.

SUNNYSIDE
Lens 1963 (Purpurine x miniature seedling) x Rosina

The flowers of yellow with a pink blush become more pink as they age. Light green foliage combines well with the flowers.

SWEDISH DOLL
Moore 1976 Fire King x Little Buckaroo

Well-shaped buds of coral red open to flat semi-single flowers showing yellow stamens. The disease-free bush can grow quite large. One year we forgot to prune our plant and next season it was taller than the 1.5 m fence beside which it was planted. It is very floriferous.

SWEETHEART ROSE
See 'Cécile Brunner'.

SWEET RASPBERRY
Jolly

We are often asked for a fragrant miniature rose. We have found that any rose is more likely to have a fragrance in the warm weather. This variety is very aptly named, as it is not only raspberry red in colour, but also has a distinct raspberry fragrance.

TAKAPUNA (Mactenni)
McGredy 1978 New Penny x ([Clare Grammerstorff x Cavalcade] x Elizabeth of Glamis)

The medium-sized double flowers of peach pink are borne on a free-flowering tall spreading bush. Another of the larger-growing miniature roses.

TEA PARTY
Moore 1972 (*R. wichuraiana* x Floradora) x Eleanor

Small double flowers of orange-apricot to pink in abundance on a short bushy plant.

TEMPLE BELLS
Morey 1971 *R. wichuraiana* x Blushing Jewel

This variety deserves a place only because it is one of the earlier groundcover roses. It has seven-petalled white blooms which flower mainly in the spring. The stems are very light, very thorny and grow flat along the ground. It is quite vigorous. The small glossy foliage is quite attractive but the variety is seldom grown now as it has been replaced by others that are so much better.

TIFFIE

SWEET RASPBERRY

THE FAIRY
See page 72.

THUNDERCLOUD
Moore 1979 Little Chief x Fire Princess

With its bushy habit of growth and large heads of blooms, this variety looks like a bright orange-red version of 'My Valentine'. It gives a bright splash of colour in the garden.

TIFFIE
Bennett 1979 Little Darling x Over The Rainbow

The long slender pale apricot pink buds open to semi-single flowers which have a touch of soft yellow in the centre. In our garden, this variety is one of the earliest to flower in the spring, and every plant is smothered with blooms.

TOM THUMB
See 'Peon'.

TOY CLOWN
Moore 1966 Little Darling x Magic Wand

Small buds of white tipped with light red open to twenty-petalled flowers with a deep cerise-red edge. The effect is very striking. Unfortunately, we have found this variety to be difficult to establish in colder areas.

THUNDERCLOUD

TOY CLOWN

TWINKLE TWINKLE
Bennett 1981 Contempo x Sheri Anne

The star-shaped flowers are an attractive blend of cream, pink and a soft apricot orange. As the bloom ages the pink becomes more pronounced. The plant is bushy and has many flowers on long slender stems.

VALERIE JEANNE
Saville 1980 Sheri Anne x Tamango

The deep magenta pink flowers are a very good shape and come on a vigorous healthy bush. Because of the very definite colour, we find people either love this variety or loathe it.

WAITEMATA (Macweemat)
McGredy 1978 Wee Man x Matangi

The very free-blooming plant shows many double flowers of crimson red which open to rosette-like blooms showing many stamens. The plant is vigorous and healthy.

TWINKLE TWINKLE

VALERIE JEANNE

WAITEMATA (Macweemat)

WANAKA (Macinca)

WATERCOLOUR

WEE MAN

WANAKA (Macinca)
McGredy 1978 Anytime x
Trumpeter

Very brilliant orange red flowers
with unusual petal formation make
this variety a real attraction in the
garden. The plant is often likened
to a geranium. Moderately
compact, it has many flowers open
at one time.

WATERCOLOUR
Moore 1975 Rumba x (Little
Darling x Red Germain)

The blooms are a beautiful blend of
light and deep pink and are a good
shape. They are borne on a
vigorous upright plant with larger
foliage.

WEE MAN
McGredy 1974 Little Flirt x
Marlena

The semi-single flowers of medium
red with striking gold stamens are
borne in clusters on a healthy bush
of medium size. 'Wee Man' can be
quite spectacular in the garden,
particularly if a group are planted
together.

WHIPPED CREAM

WHIPPED CREAM
Moore 1968 (*R. wichuraiana* x
Carolyn Dean) x White King

The globular bud opens to a double
white informal flower touched with
cream in the centre. The bush can
be quite large, and we find that the
flowers repeat more slowly than
some varieties.

WHITE ANGEL
Moore 1971 (*R. wichuraiana* x
Floradora) x (Little Darling x
miniature seedling)

The dainty white hybrid tea-
shaped buds open to attractive
small flowers with pointed petals.
The short plant is compact and has
many branches and flowers.

WHITE DREAM (Lenvir)
Lens 1982

The white flowers are larger than
usual and a very good shape from
bud through to fully open. The
foliage is dark and leathery and the
plant vigorous and healthy.

WHITE ANGEL

WHITE DREAM (Lenvir)

WHITE MADONNA

YELLOW DOLL

WINDY CITY

YELLOW BANTAM

WHITE MADONNA
Moore 1973 (*R. wichuraiana* x Floradora) x (Little Darling x miniature seedling)

The white well-shaped flowers tend to go a pale pink in the cooler weather, or if there is rain. The bush is compact and the foliage is a pale green.

WINDY CITY
Moore 1974 Little Darling x (Little Darling x (*R. wichuraiana* x seedling))

The large flowers are a deep rose pink and the foliage and plant are in proportion.

WOMEN'S OWN
McGredy

The double pink flowers come singly or several together on a moderately compact plant with large foliage.

YELLOW BANTAM
Moore 1960 (*R. wichuraiana* x Floradora) x Fairy Princess

The tiny lemon buds open to double flowers which change to nearly white soon after opening. The plant is small and the foliage is fine and dark green.

YELLOW DOLL
Moore 1962 Golden Glow x Zee

The well-shaped buds open to double medium yellow flowers with light green foliage on a moderately compact plant.

YELLOW NECKLACE   ZWERGKÖNIG 78

YELLOW NECKLACE
Moore 1965 Golden Glow x Magic Wand

The double flowers are a straw yellow, and come on a plant which can sometimes lack vigour.

ZINGER
Schwartz 1978 Zorina x Magic Carrousel

The semi-double flowers of bright red with a yellow centre sometimes come singly, but more often several together. The buds are shapely. The plant grows well and is moderately compact.

ZWERGENFEE
Kordes 1979 Miniature seedling x Träumerei

We have not seen this, or the next two miniatures. They are all from Kordes and belong to their series of bright, slightly larger, bedding varieties. We have therefore used the descriptions supplied to us.

  'Zwergenfee' has filled flowers, orange-red, a compact and bushy plant and nice foliage, green, light and glossy.

ZWERGKÖNIGIN (Queen of the Dwarfs)
Kordes 1955

The deep pink flowers, always in trusses have a light scent. The plant is compact with bushy, dark green, small foliage.

ZWERGKÖNIG 78
Kordes 1978

The compact plant is very vigorous and the flowers are blood red. Very hardy, this rose is bushy with dark green foliage. Suitable also in pots, containers and for low hedges.

ZWERGKÖNIGIN (Queen of the Dwarfs)

ZINGER

ZWERGENFEE

Because there are so many different varieties of miniature roses available it can be rather difficult to select appropriate plants for special purposes. For this reason, we have included the following specific lists of miniature roses. While we would have eventually reached agreement on the best six roses in each group, we could have had problems. The solution was easy. We have included two lists — a 'his' and 'hers'. In this way you get nearly twice as many roses and the book gets finished! Any rose which appears in both lists deserves special consideration. We have deliberately included a full range of colours. It made the choice more difficult, as in some cases we would have been happier to have had another pink or bicolour in the list rather than the yellow we needed to include. But then your favourite colour may be yellow, so that rose is there just for you. Don't look for the very latest introductions; remember it is our personal selection and we have kept it to those varieties we have seen and grown for several years.

Starters

For those who have not grown miniature roses before. These we know are attractive and grow well, with few problems.

HIS	HERS
Beauty Secret	Baby Masquerade
Jet Trail	Beauty Secret
Judy Fischer	Golden Angel
Magic Carrousel	Jet Trail
Mary Marshall	Judy Fischer
Orange Honey	Mary Marshall
Rise 'n Shine	Petite Folie
Sierra Sunrise	Wee Man

Micro-minis

Smaller varieties for those who want their miniatures to be really miniature.

HIS	HERS
Baby Betsy McCall	Baby Betsy McCall
Cinderella	Bambino
Dwarfking	Claret
Little Linda	Little Linda
Midget	Stacey Sue
White Angel	White Angel

Climbers and other tall varieties

For those who want to see the flowers without having to bend over.

HIS	HERS
Freegold	Golden Song
Jeanne Lajoie	Jeanne Lajoie
Little Girl	Little Girl
Magic Dragon	Pink Cameo
Pacesetter	Pink Petticoat
Pink Petticoat	Swedish Doll
Swedish Doll	

Exhibitors

For those who wish to compete in the local rose show. Although some of these varieties may need a little special care, they all give shapely flowers which are good for exhibiting.

HIS	HERS
Baby Katie	Cécile Lens
Cécile Lens	Cupcake
Fairlane	Hula Girl
Pacesetter	Kathy Robinson
Paragon	Pacesetter
Party Girl	Paragon
Petite Folie	Peachy Keen
Rise 'n Shine	Pink Delight
Starglo	Pink Petticoat
Starina	Rise 'n Shine
Watercolour	Valerie Jean

Favourites

Our personal selection. These are the miniature roses we would want to grow if we were marooned on a desert island!

HIS	HERS
Hula Girl	Fairlane
Little Girl	Lemon Delight
Pacesetter	Little Girl
Party Girl	Pacesetter
Pink Petticoat	Party Girl
Rose Hills Red	Peaches 'n Cream
Rise 'n Shine	Pink Petticoat
Watercolour	Ragtime

CLIMATE AS A FACTOR

Weather and climate affect the growth of all plants and miniature roses are no exception. We will describe some of the significant features of the Christchurch climate so that you can compare your own local growing conditions.

Christchurch has a temperate climate and the rainfall of about 650 mm (26 ins) a year is adequate for plant growth in all but the hottest months of summer when the temperatures can rise to more than 30°C (86°F). It is these day to day combinations of temperature and rainfall which most affect plant growth.

The warmest weather occurs in the latter part of December, January and early February. At this time it is necessary to water plants in containers every day and the hose is used on the garden once, and sometimes twice, a week as there can be two or three weeks with no significant rainfall. To make conditions worse, the wind from the north-west is warm and drying and often strong and gusty. When it blows, the humidity may fall to as little as 30 per cent and at these times plants quickly dry out and containers have to be watered a second time during the day. If temperatures are high when the nor'-wester blows, even plants in the ground suffer, as leaves and petals dehydrate unless the ground has been kept damp. Provided adequate water is available, plants grow well. It is too hot and dry for greenfly, but spider mites love these conditions.

With the cooler conditions of autumn, watering becomes less essential in the garden. The miniature roses go on flowering until the first frost of May or early June. Plants growing on the warmer north-facing side of a fence or building may in fact not lose their leaves and go on flowering all the winter.

Frosts normally occur on a number of nights during the winter, but it is not cold enough for open water to remain frozen. Snow falls occasionally, but usually melts within a few hours. Miniature roses in the open go dormant and lose their leaves, but rarely suffer any damage through winter cold.

It is necessary to be aware of the rain, or lack of it, in the spring, as a spell of dry weather can adversely affect the new growth. A late frost in September or early October can damage the soft new growth on the roses, but even this does not worry the miniature roses which have been outside all the winter.

BIBLIOGRAPHY

The American Rose Society. *The American Rose Annual*, annual publication.

Buist, Robert. *The Rose Manual*, Philadelphia, 1844 (facsimile edition Heyden, London, 1978).

Fitch, Charles Marden. *The Complete Book of Miniature Roses*, Hawthorn Books, New York, 1977.

Gault, S. Millar and Synge, Patrick M. *The Dictionary of Roses*, Ebury Press, London, 1971.

Gore, Catherine Francis. *The Book of Roses or The Rose Fanciers Manual*, Colburn, London, 1838 (facsimile edition Heyden, London, 1978).

Griffiths, Trevor. *My World of Old Roses*, Whitcoulls, Christchurch, 1983.

Hay, D. and Son, Nurserymen. *Catalogue*, Parnell, Auckland, 1899-1900.

Keays, Ethelyn Emery. *Old Roses*, Macmillan, New York, 1935 (facsimile edition Heyden, London, 1978).

Meikle, Catherine E. (ed.). *Modern Roses 8*, Mcfarland, Pennsylvania, 1980.

Moore, Ralph S. *The Story of Moore Miniature Roses*, Moore-Sequoia, Visalia, California, 1974.

The New Zealand Rose Society. *The New Zealand Rose Annual*, annual publication.

Paterson, Allen. *The History of the Rose*, Collins, London, 1983.

Paul, William. *The Rose Garden*, Sherwood, Gilbert & Piper, London, 1848 (facsimile edition Heyden, London, 1978).

Pinney, Margaret E. *The Miniature Rose Book*, Van Nostrand, Princeton, New Jersey, 1964.

The Royal National Rose Society. *The Rose Annual*, annual publication.

Shepherd, Roy E. *History of the Rose*, Macmillan, 1954 (facsimile edition Heyden, London, 1978).

The Standard Cyclopedia of Horticulture, New York, 1917.

Wylie, A. P. 'The History of Garden Roses' in *Journal of the Royal Horticultural Society*, Vol. 79, 1954.

Young, Norman. *The Complete Rosarian*, L. A. Wyatt (ed.), Hodder & Stoughton, London, 1971.

INDEX